Anderson and Nash
LINEAR PROGRAMMING IN INFINITE-DIMENSIONAL SPACES: THEORY AND APPLICATION

Minc
NONNEGATIVE MATRICES

Ahlswede and Wegener
SEARCH PROBLEMS

Beale
INTRODUCTION TO OPTIMIZATION

Nonnegative Matrices

Nonnegative Matrices

Henryk Minc
Department of Mathematics
University of California
Santa Barbara, California 93106

A WILEY-INTERSCIENCE PUBLICATION
JOHN WILEY & SONS
New York · Chichester · Brisbane · Toronto · Singapore

Library of Congress Cataloging in Publication Data:

Minc, Henryk.
Nonnegative matrices.

"A Wiley-Interscience Publication."
Bibliography: p.
Includes index.
1. Non-negative matrices. I. Title.

QA188.M558 1988 512.9′434 87-27416
ISBN 0-471-83966-3

Printed in the United States of America

10 9 8 7 6 5 4 3 2 1

To my grandson Jeffrey

Preface

In 1907 Perron discovered some remarkable properties of square matrices with positive entries. This work was substantially generalized by Frobenius who extended Perron's results to nonnegative matrices, that is, matrices with nonnegative entries. Since then the theory of nonnegative matrices has been one of the most active areas of linear algebra. It has found many applications in various parts of mathematics and in the physical and social sciences. Indeed in many universities the theory of nonnegative matrices has become a standard part of the curriculum.

This book is an outgrowth of courses that I have given over the years at the University of California at Santa Barbara and at the Technion - Israel Institute of Technology in Haifa. It is nigh impossible to write a comprehensive treatise on nonnegative matrices. The subject has become simply too extensive. My purpose in writing the book is twofold:

1. To provide a textbook for two-quarter or one-semester courses either at the undergraduate upper-division or the graduate level.
2. To write a self-contained reference work for mathematicians and scientists interested in the theory of nonnegative matrices.

Thus the problem section at the end of each chapter may be regarded as being primarily a part of the textbook, whereas the detailed references are mainly intended for advanced students and research workers. Nonetheless, the only prerequisite for the book is proficiency in matrix algebra at the level of a standard undergraduate course.

The book consists of seven chapters. Chapters I, II, and III contain the basic Perron-Frobenius theory, somewhat enlarged and brought up to date. Most of the proofs in the book are rather simpler than the original proofs of Frobenius. The first three chapters, with the possible exclusion of Sections 2.1 and 2.3, form the essential and indispensable core of any course on nonnegative matrices. The remaining four chapters depend on this core material, but are otherwise mostly self-contained. Chapter IV deals with combinatorial properties of nonnegative matrices. In Chapter V the theory of doubly stochastic matrices is presented in detail. The concluding section of the chapter contains a complete proof of the van der Waerden permanent conjecture. Chapter VI deals with important special classes of nonnegative matrices which find many applications in applied sciences. The last chapter presents problems of existence of nonnegative matrices, stochastic and doubly stochastic matrices, with prescribed eigenvalues or elementary divisors. Most of these inverse eigenvalue problems are unsolved.

It has become fashionable to include in mathematical textbooks all kind of "applications" to as many cognate and unrelated fields as possible. Thus a textbook on our topic could include chapters or sections dealing with applications of nonnegative matrices in probability, combinatorics, numerical analysis, dynamic programming, operations research, physics, chemistry, economics, sociology, demography, and in other disciplines in which numerical models can be represented by nonnegative matrices. However, most of such applications would be, by necessity, of infinitesimal depth, and any particular application could interest but a small number of readers. Indeed, the only interest they all may have in common is the mathematical theory of nonnegative matrices. The aim of this book is to provide the reader with a rigorous study of the Perron-Frobenius theory and some of its more recent developments and outgrowths. It is hoped that it will be helpful to anyone interested in this important area of linear algebra, whether in its theoretical aspects or in its applications.

My thanks are due to Dr. Arnold R. Kräuter and Mr. Hervé Moulin, who read the manuscript and helped to eliminate many errors and obscurities. The work on the book was supported in part by the Office of Naval Research under Research Contract N00014-85-K-0489.

HENRYK MINC

Santa Barbara, 1987

Contents

Notation and Terminology

The notation and the terminology used in this book are essentially the same as in *Survey of Matrix Theory and Matrix Inequalities* by M. Marcus and H. Minc, *Permanents* by H. Minc, and in many other books on linear algebra and matrix theory. For convenience, we give below a list of the more important symbols and their definitions. Many of them are also defined in the text. An index of symbols used in the book is to be found on page 197.

1. *General Matrices.* Let $\mathbb{R}$, $\mathbb{C}$, and $\mathbb{P}$ denote the set of real numbers, complex numbers, and nonnegative numbers, respectively. If S is any set, then $M_{m,n}(S)$ represents the set of $m \times n$ matrices with entries from the set S. If $m = n$, we use the abbreviated notation $M_n(S)$.

Matrices are represented by capital italic Latin letters, and their entries usually by lowercase italic Latin letters. The statement "$A = (a_{ij})$" means that A is the matrix whose (i, j) entry is denoted by a_{ij}. The (i, j) entry of a matrix A is also denoted by A_{ij}, although in the context of partitioned matrices A_{ij} may represent the (i, j) block of A.

The ith row and the jth column of a matrix A are denoted by $A_{(i)}$ and $A^{(j)}$, respectively. A *line* of A designates either a row or a column of A.

The $n \times n$ identity matrix is denoted by I_n, or sometimes just by I. The $m \times n$ zero matrix is denoted by $0_{m,n}$, or simply by 0. The diagonal matrix with $d_1, d_2, \ldots, d_n$ as its main diagonal entries is designated by $\operatorname{diag}(d_1, d_2, \ldots, d_n)$. The $n \times n$ matrix all of whose entries are $1/n$ is denoted by J_n. The symbol J without a subscript represents a matrix, of appropriate order, whose entries are all equal to 1. The matrix whose (i, j) entry is 1 and all other entries are 0 is denoted by E_{ij}. In general, a matrix all of whose entries are 0's and 1's is called a *(0, 1)-matrix.*

If $A = (a_{ij}) \in M_{m,n}(S)$, then $A^{\mathrm{T}} \in M_{n,m}$ designates the *transpose* of A. The symbols $|A|$ and $\bar{A}$ denote the $m \times n$ matrices whose (i, j) entries are $|a_{ij}|$ and $\bar{a}_{ij}$, respectively; $\bar{A}$ is called the *conjugate* of A. The *adjoint* (that is, the conjugate transpose) of A is denoted by A^*.

The symbols $\Sigma^{\boldsymbol{\cdot}}$ and $\dotplus$ are used to represent direct sums. The symbol $*$ represents the *Hadamard product*: If $A = (a_{ij})$ and $B = (b_{ij})$ are $m \times n$ matrices, then $A * B$ is the $m \times n$ matrix whose (i, j) entry is $a_{ij}b_{ij}$ for all i and j.

2. *Scalar-Valued Functions of Matrices.* The determinant of a square matrix A is denoted by $\det(A)$, or simply by $\det A$. The *permanent* of an $m \times n$ matrix $A = (a_{ij})$, $m \leq n$, is denoted by $\operatorname{Per}(A)$, or $\operatorname{Per} A$; it is defined by

$$\operatorname{Per}(A) = \sum_{\sigma} \prod_{i=1}^{m} a_{i\sigma(i)},$$

where the summation extends over all one–one functions from $\{1, 2, \ldots, m\}$ to $\{1, 2, \ldots, n\}$. If $m = n$, then the permanent is denoted by $\operatorname{per}(A)$, or $\operatorname{per} A$.

The rank of matrix A is denoted by $\rho(A)$, and its trace by $\operatorname{tr}(A)$.

3. *Index Sets and Submatrices.* If k and n are integers, $1 \leq k \leq n$, then $Q_{k,n}$ and $G_{k,n}$ denote the set of increasing sequences of integers,

$$\omega = (\omega_1, \omega_2, \ldots, \omega_k), \qquad 1 \leq \omega_1 < \omega_2 < \cdots < \omega_k \leq n,$$

and the set of nondecreasing sequences of integers,

$$\omega = (\omega_1, \omega_2, \ldots, \omega_k), \qquad 1 \leq \omega_1 \leq \omega_2 \leq \cdots \leq \omega_k \leq n,$$

respectively. If $\alpha \in G_{k,n}$, then $\mu(\alpha)$ represents the product of the factorials of the multiplicities of the distinct integers appearing in the sequence α. If $\alpha = (\alpha_1, \alpha_2, \ldots, \alpha_m)$ is a sequence of integers and h is an integer, then $\alpha + h$ and (α, h) represent the sequences $(\alpha_1 + h, \alpha_2 + h, \ldots, \alpha_m + h)$ and $(\alpha_1, \alpha_2, \ldots, \alpha_m, h)$, respectively. If $\alpha \in Q_{k,n}$ and p, q are integers, $1 \leq p < q \leq k$, then $\alpha^{(p,q)}$ denotes the subsequence $(\alpha_p, \alpha_{p+1}, \ldots, \alpha_q)$.

Let $A = (a_{ij}) \in M_{m,n}(S)$, and let $\alpha = (\alpha_1, \alpha_2, \ldots, \alpha_h)$ and $\beta = (\beta_1, \beta_2, \ldots, \beta_k)$ be sequences in $Q_{h,m}$ and $Q_{k,n}$, respectively. Then

$$A[\alpha|\beta] = A[\alpha_1, \alpha_2, \ldots, \alpha_h|\beta_1, \beta_2, \ldots, \beta_k]$$

denotes the $h \times k$ submatrix of A whose (i, j) entry is $a_{\alpha_i\beta_j}$, $i = 1, 2, \ldots, h$, $j = 1, 2, \ldots, k$; and

$$A(\alpha|\beta) = A(\alpha_1, \alpha_2, \ldots, \alpha_h|\beta_1, \beta_2, \ldots, \beta_k),$$

is the $(m - h) \times (n - k)$ submatrix of A obtained from A by deleting rows $\alpha_1, \alpha_2, \ldots, \alpha_h$ and columns $\beta_1, \beta_2, \ldots, \beta_k$. Occasionally, we will use an abbreviated notation for the principal submatrices of A: Instead of $A[\alpha_1, \alpha_2, \ldots, \alpha_k|\alpha_1, \alpha_2, \ldots, \alpha_k]$ and $A(\alpha_1, \alpha_2, \ldots, \alpha_k|\alpha_1, \alpha_2, \ldots, \alpha_k)$, we write $A[\alpha_1, \alpha_2, \ldots, \alpha_k]$ and $A(\alpha_1, \alpha_2, \ldots, \alpha_k)$, respectively. In a similar fashion, $A[\alpha|\alpha]$ and $A(\alpha|\alpha)$ are abbreviated to $A[\alpha]$ and $A(\alpha)$, respectively.

The *adjugate* of an $n \times n$ matrix A, denoted by $\operatorname{adj} A$, (sometimes incorrectly called the *adjoint* of A, or the *classical adjoint* of A) is the $n \times n$ matrix whose (i, j) entry is $(-1)^{i+j}\det(A(j|i)$, $i, j = 1, 2, \ldots, n$. The rth (determinantal) *compound* of A, denoted by $C_r(A)$, is the $\binom{n}{r} \times \binom{n}{r}$ matrix whose entries are $\det(A[\alpha|\beta])$, $\alpha, \beta \in Q_{r,n}$, arranged lexicographically in α and β.

The rth *permanental compound* of A, designated by $L_r(A)$, is the $\binom{n}{r} \times \binom{n}{r}$ matrix whose entries are $\text{per}(A[\alpha|\beta])$, $\alpha, \beta \in Q_{r,n}$, arranged lexicographically in α and β. The rth *induced matrix* of A, denoted by $P_r(A)$, is the $\binom{n+r-1}{r} \times \binom{n+r-1}{r}$ matrix whose entries are $\text{per}(A[\alpha|\beta])/\sqrt{\mu(\alpha)\mu(\beta)}$, arranged lexicographically in α and β, where $\alpha, \beta \in G_{r,n}$, and μ is the function defined on page xii.

4. *Vectors*. The vector space spanned by vectors $v_1, v_2, \ldots, v_m$ is represented by $\langle v_1, v_2, \ldots, v_m \rangle$. If u and v are vectors in a unitary space, then their inner product is denoted by (u, v); the length of u is denoted by $\|u\|$. The space of n-tuples with entries (coordinates) from a set S is denoted by S^n. The special symbol E^n denotes the subset of $\mathbb{P}^n$ defined by

$$E^n = \left\{ (x_1, x_2, \ldots, x_n) \in \mathbb{P}^n \,\middle|\, \sum_{i=1}^{n} x_i = 1 \right\}.$$

If F is a field, then the vectors of the standard basis of F^n are denoted by $e_1, e_2, \ldots, e_n$.

5. *Nonnegative Matrices and Vectors*. A real matrix A is called *positive* if all its entries are positive. It is called *nonnegative* if all its entries are nonnegative. We write $A > 0$ if A is positive, and $A \geq 0$ if it is nonnegative. Similarly, a real n-tuple $x = (x_1, x_2, \ldots, x_n)$ is *positive* (*nonnegative*) if $x_i > 0$ ($x_i \geq 0$), $i = 1, 2, \ldots, n$; we write $x > 0$ ($x \geq 0$). The set of $n \times n$ doubly stochastic matrices is denoted by Ω_n. The set of $n \times n$ (0,1)-matrices with k 1's in each line is denoted by Λ_n^k. Two $m \times n$ matrices, $A = (a_{ij})$ and $B = (b_{ij})$, are said to have the same *zero pattern* if $a_{ij} = 0$ whenever $b_{ij} = 0$, and vice versa. A matrix that has the same zero pattern as a permutation matrix is called a *generalized permutation matrix*.

If α and β are nonnegative n-tuples, and α is majorized by β (see Section 5.2), then we write $\alpha \prec \beta$.

6. *Miscellaneous*. The set of all permutations of $(1, 2, \ldots, n)$ is denoted by S_n.

The *Kronecker delta* is defined by

$$\delta_{ij} = \begin{cases} 1, & \text{if } i = j, \\ 0, & \text{if } i \neq j. \end{cases}$$

If A, B, C are points in a plane, then $\Delta(A, B, C)$ denotes the triangle with A, B, C as its vertices, and $\angle(A, B, C)$ represents both the angle with B as its vertex and its measure.

I

Spectral Properties of Nonnegative Matrices

1.1. LINEAR TRANSFORMATIONS ON NONNEGATIVE MATRICES

The usual method for studying invariants defined on matrices is to simplify the structure of the matrices by linear transformations which preserve the invariants. In this chapter we are interested in spectral properties of nonnegative matrices, and the relevant question to ask is: What linear transformations on nonnegative matrices can be used in order to simplify their structure? Specifically: What linear transformations map nonnegative matrices into nonnegative matrices and hold their spectra fixed?

We first state without proof a classical result of Frobenius on linear transformations that preserve the determinant function.

Theorem 1.1 (Frobenius [2]). *If T is a linear transformation on $M_n(\mathbb{C})$ that holds the determinant of each matrix fixed, then there exist matrices U and V such that* $\det(UV) = 1$, *and*

$$T(A) = UAV,$$

for all $A \in M_n(\mathbb{C})$, or

$$T(A) = UA^{\mathrm{T}}V,$$

for all $A \in M_n(\mathbb{C})$.

Next we establish two results [7] about linear transformations on complex matrices.

Theorem 1.2. *If T is a linear transformation on $M_n(\mathbb{C})$, and T preserves the determinant and the trace of each matrix, that is,* $\det(T(A)) = \det(A)$ *and*

$\mathrm{tr}(T(A)) = \mathrm{tr}(A)$ *for all* $A \in M_n(\mathbb{C})$, *then there exists a matrix* V *in* $M_n(\mathbb{C})$ *such that*

$$T(A) = V^{-1}AV,$$

for all $A \in M_n(\mathbb{C})$, *or*

$$T(A) = V^{-1}A^{\mathrm{T}}V,$$

for all $A \in M_n(\mathbb{C})$.

Proof. Since T preserves determinants, there exist, by Theorem 1.1, matrices $U = (u_{ij})$ and $V = (v_{ij})$ such that $\det(UV) = 1$, and either

$$T(A) = UAV, \tag{1}$$

for all $A \in M_n(\mathbb{C})$, or

$$T(A) = UA^{\mathrm{T}}V, \tag{2}$$

for all $A \in M_n(\mathbb{C})$. If T is of the form (1), then

$$\begin{aligned} T(E_{ij}) &= UE_{ij}V \\ &= U^{(i)}V_{(j)}. \end{aligned}$$

Now,

$$\mathrm{tr}(E_{ij}) = \delta_{ij},$$

and

$$\begin{aligned} \mathrm{tr}(T(E_{ij})) &= \mathrm{tr}(UE_{ij}V) \\ &= \sum_{s=1}^{n} u_{si}v_{js} \\ &= (VU)_{ji}. \end{aligned}$$

Since $\mathrm{tr}(T(E_{ij})) = \mathrm{tr}(E_{ij})$, we have

$$(VU)_{ji} = \delta_{ij},$$

for all i, j, and therefore

$$VU = I_n,$$

that is,

$$T(A) = V^{-1}AV.$$

If $T(A) = UA^{T}V$ for all $A \in M_n(\mathbb{C})$, we can prove in a similar fashion that

$$\operatorname{tr}\big(T(E_{ij})\big) = (VU)_{ij},$$

for all i, j, and therefore

$$VU = I_n.$$

Hence in this case

$$T(A) = V^{-1}A^{T}V,$$

for all $A \in M_n(\mathbb{C})$. ■

Corollary 1.1. *A linear transformation on the space of complex $n \times n$ matrices holds the spectrum of each matrix fixed if and only if it preserves the trace and the determinant of each matrix.*

For the proof of the main theorem of this section we require the following lemma.

Lemma 1.1. *The inverse of a nonnegative matrix A is nonnegative if and only if A is a generalized permutation matrix.*

Proof. The sufficiency of the condition is obvious. Let $A = (a_{ij}) \in M_n(P)$, and suppose that $A^{-1} = (b_{ij})$ is nonnegative. Thus

$$\sum_{t=1}^{n} a_{it}b_{tj} = \delta_{ij}, \qquad i, j = 1, 2, \ldots, n.$$

If the ith row of A has exactly k positive entries in positions (i, j_s), $s = 1, 2, \ldots, k$, and $j \neq i$, then b_{tj} must vanish for $t = j_s$, $s = 1, 2, \ldots, k$. In other words, A^{-1} must contain a $k \times (n-1)$ zero submatrix. But if k were greater than 1, then clearly the determinant of A^{-1} would be zero. Hence k cannot exceed 1. It follows that A has at most one positive entry in each row, and since it is nonsingular, it must be a generalized permutation matrix. ■

Theorem 1.3 (Minc [7]). *If T is a linear transformation on $M_n(\mathbb{C})$ that maps nonnegative matrices into nonnegative matrices and preserves the spectrum of each nonnegative matrix, then there exists a nonnegative generalized permutation matrix P such that*

$$T(A) = P^{-1}AP, \tag{3}$$

for all $A \in M_n(\mathbb{C})$, or

$$T(A) = P^{-1}A^{T}P, \tag{4}$$

for all $A \in M_n(\mathbb{C})$.

Proof. Since T preserves the spectra of all nonnegative matrices, it preserves their traces and determinants. Clearly, any linear transformation that preserves the trace of each nonnegative $n \times n$ matrix, and in particular of each E_{ij}, will preserve the traces of all matrices in $M_n(\mathbb{C})$. We show that T also holds the determinant of each complex matrix fixed.

Consider an $n \times n$ matrix $X = (x_{ij})$, where the x_{ij} are independent indeterminates over $\mathbb{C}$. The entries in $T(X)$ are fixed linear combinations of the entries in X. Hence $\det(T(X)) - \det(X)$ is a polynomial in the indeterminates x_{ij}. This polynomial vanishes if X is replaced by any nonnegative $n \times n$ matrix. It follows (see Problem 9) that the polynomial $\det(T(X)) - \det(X)$ is identically zero, that is,

$$\det(T(X)) = \det(X),$$

and thus

$$\det(T(A)) = \det(A),$$

for all A in $M_n(\mathbb{C})$.

Hence T preserves both the trace and the determinant of each complex $n \times n$ matrix and therefore, by Theorem 1.2, there exists a matrix S such that

$$T(A) = S^{-1}AS, \tag{5}$$

for all $A \in M_n(\mathbb{C})$, or

$$T(A) = S^{-1}A^{\mathrm{T}}S, \tag{6}$$

for all $A \in M_n(\mathbb{C})$. If T is of the form (5), then

$$\begin{aligned} T(E_{ij}) &= S^{-1}E_{ij}S \\ &= (S^{-1})^{(i)}S_{(j)} \end{aligned}$$

is nonnegative for all i and j. Hence

$$(S^{-1})_{hi}S_{jk} \geq 0,$$

for all h, i, j, and k. Now, not all $(S^{-1})_{hi}$ can be zero. Hence there is a complex number α such that

$$S_{jk} = \alpha|S_{jk}|,$$

for all j and k. In other words, S is a scalar multiple of a nonnegative matrix P, and therefore

$$S^{-1}AS = P^{-1}AP,$$

for all A. We have

$$\begin{aligned} T(E_{ij}) &= P^{-1}E_{ij}P \\ &= (P^{-1})^{(i)}P_{(j)}, \end{aligned}$$

which must be nonnegative for all i and j. It follows, as before, that

$$(P^{-1})_{hi}P_{jk} \geq 0,$$

for all h, i, j, and k. Since some of the P_{jk} must be positive, all the $(P^{-1})_{hi}$ must be nonnegative, that is, P^{-1} must be a nonnegative matrix. The result follows by Lemma 1.1. If T is of the form (6), the proof is similar. ■

It is clear from the first part of the proof and Corollary 1.1 that the apparently weaker premise, that T is a linear transformation on nonnegative matrices that merely holds the determinant and the trace fixed, is sufficient to imply the conclusion of Theorem 1.3.

1.2. IRREDUCIBLE MATRICES

A matrix X is said to be *cogredient* to a matrix Y if there exists a permutation matrix P such that $X = P^{\mathrm{T}}YP$.

Definition 2.1. A nonnegative n-square matrix A, $n \geq 2$, is called *reducible* (*decomposable*) if it is cogredient to a matrix of the form

$$\begin{bmatrix} B & C \\ 0 & D \end{bmatrix},$$

where B and D are square submatrices. Otherwise, A is *irreducible* (*indecomposable*). Clearly, A is reducible if and only if there exists an ordering $(i_1, i_2, \ldots, i_s, j_{s+1}, j_{s+2}, \ldots, j_n)$ of $(1, 2, \ldots, n)$ such that $A[i_1, i_2, \ldots, i_s | j_{s+1}, j_{s+2}, \ldots, j_n] = 0$. A 1×1 matrix is irreducible, by definition.

Example 2.1. Show that if $A = (a_{ij}) \geq 0$ is an irreducible $n \times n$ matrix and $x = (x_1, x_2, \ldots, x_n) \geq 0$, then $Ax = 0$ implies that $x = 0$.

Suppose that $x_k > 0$. Then

$$\begin{aligned} 0 = (Ax)_i &= \sum_{j=1}^{n} a_{ij}x_j \\ &\geq a_{ik}x_k, \end{aligned}$$

and therefore $a_{ik} = 0$, for all i. In other words, A has a zero column and thus is reducible. ■

We next prove an important property of irreducible matrices.

Theorem 2.1. *If A is an irreducible nonnegative $n \times n$ matrix, $n \geq 2$, and y is a nonnegative n-tuple with exactly k positive coordinates, $1 \leq k \leq n-1$, then $(I_n + A)y$ has more than k positive coordinates.*

Proof. Suppose that k coordinates of y are positive and the others are zero. Let P be a permutation matrix such that the first k coordinates of $x = Py$ are positive and the others are zero. Since $A \geq 0$, the number of zero coordinates in $(I_n + A)y = y + Ay$ cannot be greater than $n - k$. Suppose it is $n - k$. This would mean that $(Ay)_i = 0$ whenever $y_i = 0$, that is, $(PAy)_i = 0$ whenever $(Py)_i = 0$. But $Py = x$, and therefore the assumption that $(I_n + A)y$ has as many 0's as y is equivalent to the assertion that $(PAP^{\mathrm{T}}x)_i = 0$ for $i = k+1, k+2, \ldots, n$. Let $B = (b_{ij}) = PAP^{\mathrm{T}}$. Then

$$\begin{aligned} (Bx)_i &= \sum_{j=1}^{n} b_{ij}x_j \\ &= \sum_{j=1}^{k} b_{ij}x_j \\ &= 0, \end{aligned}$$

for $i = k+1, k+2, \ldots, n$. But $x_j > 0$ for $1 \leq j \leq k$ and therefore $b_{ij} = 0$ for $i = k+1, k+2, \ldots, n$ and $j = 1, 2, \ldots, k$. Thus if $(I_n + A)y$ had the same number of zero coordinates as y, the matrix A would have to be reducible. ■

Corollary 2.1. *If A is an irreducible $n \times n$ matrix, and y is a nonzero nonnegative n-tuple, then $(I_n + A)^{n-1}y > 0$.*

Corollary 2.2. *An n-square nonnegative matrix A is irreducible if and only if $(I_n + A)^{n-1} > 0$.*

Proof. If A is irreducible, then

$$(I_n + A)^{n-1}e_j > 0,$$

for $j = 1, 2, \ldots, n$. In other words, all the columns of $(I_n + A)^{n-1}$ are positive. The converse is obvious [see Problems 3(c) and 3(j)]. ■

Theorem 2.2. *A nonnegative eigenvector of a nonnegative irreducible matrix must be strictly positive.*

Proof. Suppose that

$$Ax = \lambda x,$$

where $A \geq 0$ is irreducible, $x \geq 0$, and $x \neq 0$. Clearly, λ must be nonnegative. Now,

$$(I_n + A)x = (1 + \lambda)x.$$

If x had k zero coordinates, $1 \leq k < n$, then $(1 + \lambda)x$ would have k zeros as well, whereas, by Theorem 2.1, $(I_n + A)x$ would have less than k zeros. Hence x must be positive. ■

Let $a_{ij}^{(k)}$ denote the (i, j) entry of A^k, the kth power of $A = (a_{ij})$.

Theorem 2.3. *A nonnegative square matrix $A = (a_{ij})$ is irreducible if and only if for each (i, j) there exists an integer k such that $a_{ij}^{(k)} > 0$.*

Proof. Suppose that A is irreducible. Then, by Corollary 2.2,

$$(I_n + A)^{n-1} > 0.$$

Let $B = (b_{ij}) = (I_n + A)^{n-1}A$. Clearly, $B > 0$. Let

$$B = A^n + c_{n-1}A^{n-1} + \cdots + c_2A^2 + c_1A.$$

Then

$$b_{ij} = a_{ij}^{(n)} + c_{n-1}a_{ij}^{(n-1)} + \cdots + c_2a_{ij}^{(2)} + c_1a_{ij} > 0,$$

for all (i, j). It follows that for each (i, j) there must exist an integer k, $1 \leq k \leq n$, such that $a_{ij}^{(k)} > 0$.

To prove the converse we have to show that if A is reducible, then $a_{ij}^{(k)} = 0$ for some (i, j), whatever the integer k. Suppose that A is a reducible $n \times n$ matrix and that P is a permutation matrix such that

$$P^{\mathrm{T}}AP = \begin{bmatrix} B & C \\ 0 & D \end{bmatrix},$$

where B is $s \times s$. But then for all i and j satisfying $s + 1 \leq i \leq n$ and $1 \leq j \leq s$, the (i, j) entry of $P^{\mathrm{T}}A^kP$ is zero for any k. ■

1.3. THE COLLATZ–WIELANDT FUNCTION

Let S be a subset of the set of complex numbers. A problem of considerable interest in linear algebra is to determine how the spectral properties of a matrix are affected by restricting its entries to S. For example, if S is the set of real numbers, then the spectra of matrices over S are symmetric relative to the real axis. If S is the set of algebraic numbers, then all the eigenvalues of any matrix over S are algebraic. However, in general, such problems appear to be either unsolvable or perhaps without a clearcut solution. In 1907 Perron [8] discovered some remarkable and unexpected spectral properties of positive matrices. Frobenius [3, 4] extended and greatly amplified Perron's results by generalizing them to irreducible nonnegative matrices. Various proofs of the *Perron–Frobenius theory* have appeared in the literature. The methods of proof in this book follow in most cases the elegant method developed by Wielandt in [9].

Let E^n be the subset of $\mathbb{P}^n$ defined by

$$E^n = \left\{ (x_1, x_2, \ldots, x_n) \in \mathbb{P}^n \,\middle|\, \sum_{i=1}^{n} x_i = 1 \right\}.$$

Definition 3.1. Let $A = (a_{ij})$ be an irreducible $n \times n$ nonnegative matrix. Define the function f_A from $\mathbb{P}^n$ to the set of nonnegative numbers by

$$f_A(x) = \min_{x_i \neq 0} \frac{(Ax)_i}{x_i},$$

for all nonzero $x = (x_1, x_2, \ldots, x_n) \in \mathbb{P}^n$. The function f_A is called the *Collatz–Wielandt function* associated with A [1, 9].

Theorem 3.1. *Let A be an irreducible nonnegative matrix and let f_A be the Collatz–Wielandt function associated with A. Then*

(i) *the function f_A is homogeneous of degree* 0;

(ii) *if x is nonnegative, nonzero, and ρ is the largest real number for which*

$$Ax - \rho x \geq 0,$$

then $\rho = f_A(x)$;

(iii) *if $x \in \mathbb{P}^n$, $x \neq 0$, and $y = (I_n + A)^{n-1}x$, then $f_A(y) \geq f_A(x)$.*

Proof. (i) For $t > 0$ and $x \in \mathbb{P}^n$, $x \neq 0$, we have

$$\begin{aligned} f_A(tx) &= \min_{(tx)_i \neq 0} \frac{(A(tx))_i}{(tx)_i} \\ &= \min_{tx_i \neq 0} \frac{t(Ax)_i}{tx_i} \\ &= \min_{x_i \neq 0} \frac{(Ax)_i}{x_i} \\ &= f_A(x). \end{aligned}$$

(ii) The definition of f_A implies that

$$Ax - f_A(x)x \geq 0,$$

and that there exists an integer k, $1 \leq k \leq n$, such that $x_k \neq 0$ and the kth coordinate of $Ax - f_A(x)x$ is 0. Thus if $c > f_A(x)$, then the kth coordinate of $Ax - cx$ is negative. The result follows.

(iii) We have

$$Ax - f_A(x)x \geq 0.$$

Multiplying both sides by $(I_n + A)^{n-1}$, we obtain

$$A(I_n + A)^{n-1}x - f_A(x)(I_n + A)^{n-1}x \geq 0,$$

since A and $(I_n + A)^{n-1}$ commute, that is,

$$Ay - f_A(x)y \geq 0.$$

But by part (ii), $f_A(y)$ is the largest number satisfying

$$Ay - f_A(y)y \geq 0.$$

Hence

$$f_A(y) \geq f_A(x). \quad \blacksquare$$

Example 3.1. Let $A = (a_{ij})$ be a nonnegative irreducible $n \times n$ matrix. Show that the function f_A is bounded.

Clearly, f_A is bounded below by 0. We show that it is bounded above by the largest column sum of A. Let

$$c_j = \sum_{i=1}^{n} a_{ij}, \qquad j = 1, 2, \ldots, n.$$

In view of Theorem 3.1(i) it suffices to prove that

$$f_A(x) \leq \max_j c_j,$$

for any $x \in E^n$. Now,

$$(Ax)_i \geq f_A(x)x_i,$$

that is,

$$\sum_{j=1}^{n} a_{ij}x_j \geq f_A(x)x_i,$$

for $i = 1, 2, \ldots, n$. Therefore summing with respect to i, we have

$$\sum_{i=1}^{n} \sum_{j=1}^{n} a_{ij} x_j \geq \sum_{i=1}^{n} f_A(x) x_i$$
$$= f_A(x),$$

since $\sum_{i=1}^{n} x_i = 1$. On the other hand,

$$\sum_{i=1}^{n} \sum_{j=1}^{n} a_{ij} x_j = \sum_{j=1}^{n} x_j \sum_{i=1}^{n} a_{ij}$$
$$= \sum_{j=1}^{n} x_j c_j$$
$$\leq \max_j c_j,$$

and therefore

$$f_A(x) \leq \max_j c_j. \quad \blacksquare$$

Theorem 3.2. *Let A be an irreducible nonnegative $n \times n$ matrix. Then the function f_A attains its maximum in E^n.*

Note that E^n is closed and bounded and thus it is compact. If the function f_A were continuous on E^n, the result would follow immediately. Clearly, f_A is continuous at any positive n-tuple in E^n. However, the function f_A may not be continuous on the boundary of E^n. For example, if

$$A = \begin{bmatrix} 2 & 2 & 1 \\ 2 & 2 & 1 \\ 0 & 2 & 1 \end{bmatrix}$$

and $x(\varepsilon) = (1, 0, \varepsilon)/(1 + \varepsilon)$, where $\varepsilon > 0$, then

$$Ax(\varepsilon) = (2 + \varepsilon, 2 + \varepsilon, \varepsilon)/(1 + \varepsilon),$$

and

$$f_A(x(\varepsilon)) = \min\left(\frac{2+\varepsilon}{1}, \frac{\varepsilon}{\varepsilon}\right)$$
$$= 1.$$

However,

$$f_A(x(0)) = 2 \neq 1 = \lim_{\varepsilon \to 0} f_A(x(\varepsilon)).$$

Proof of Theorem 3.2. Let

$$G = (I_n + A)^{n-1}E^n = \left\{ y \middle| y = (I_n + A)^{n-1}x,\ x \in E^n \right\}.$$

Then G is a compact set. Also, by Corollary 2.1, all the n-tuples in G are strictly positive. Hence f_A is continuous on G. Since G is compact, the function f_A attains its maximum value in G at some $y^0 = (y_1^0, y_2^0, \ldots, y_n^0) \in G$. Let $x^0 = y^0/\sum_{i=1}^n y_i^0 \in E^n$, and let x be any vector in E^n. Then if $y = (I_n + A)^{n-1}x$, we have

$$\begin{aligned} f_A(x) &\le f_A(y), && \text{by Theorem 3.1(iii)}, \\ &\le f_A(y^0), && \text{by the maximality of } y^0 \text{ in } G, \\ &= f_A(x^0), && \text{by Theorem 3.1(i)}. \end{aligned}$$

Since x was any vector in E^n, it follows that f_A has an absolute maximum in E^n at x^0. ■

1.4. MAXIMAL EIGENVALUE OF A NONNEGATIVE MATRIX

The following theorem is the best known and perhaps the most important part of the Perron–Frobenius theory.

Theorem 4.1. *An irreducible nonnegative matrix A has a real positive eigenvalue r such that*

$$r \ge |\lambda_i|,$$

for any eigenvalue λ_i of A. Furthermore, there is a positive eigenvector corresponding to r.

(The eigenvalue r is called the *maximal eigenvalue* of A, and a positive eigenvector corresponding to r is called a *maximal eigenvector* of A.)

Proof. Let A be an irreducible nonnegative $n \times n$ matrix. By Theorem 3.2, there exists a vector x^0 in E^n such that

$$f_A(x^0) \ge f_A(x),$$

for all x in E^n. Let

$$r = f_A(x^0),$$

that is,

$$r = \max\{ f_A(x) | x \in E^n \}.$$

We first show that r is positive. Let $u = (1, 1, \ldots, 1)/n$. Then

$$\begin{aligned} r &\geq f_A(u) \\ &= \min_i \frac{(Au)_i}{u_i} \\ &= \min_i \sum_{j=1}^{n} a_{ij} \\ &> 0, \end{aligned}$$

since A cannot have a zero row. Next, we show that r is an eigenvalue of A. We certainly have

$$Ax^0 - rx^0 \geq 0. \tag{1}$$

Suppose that $Ax^0 - rx^0 \neq 0$. Then, by Corollary 2.1,

$$(I_n + A)^{n-1}(Ax^0 - rx^0) > 0,$$

that is,

$$Ay^0 - ry^0 > 0, \tag{2}$$

where $y^0 = (I_n + A)^{n-1}x^0$. Since (2) is a strict inequality there exists a positive number ε such that

$$Ay^0 - (r + \varepsilon)y^0 \geq 0.$$

But then, by Theorem 3.1(ii),

$$r + \varepsilon \leq f_A(y^0),$$

and therefore

$$r < f_A(y^0),$$

which contradicts the maximality of r. Hence (1) is an equality, r is an eigenvalue, and x^0 is a nonnegative eigenvector corresponding to r.

Note that we have actually shown that if x is a nonnegative nonzero vector and

$$Ax - rx \geq 0,$$

then x is an eigenvector of A corresponding to r. Then Theorem 2.2 implies that $x > 0$.

Next, let $Az = \lambda_i z$ where $z = (z_1, z_2, \ldots, z_n) \neq 0$. Then

$$\lambda_i z_t = \sum_{j=1}^{n} a_{tj} z_j, \qquad t = 1, 2, \ldots, n,$$

and therefore

$$|\lambda_i||z_t| \le \sum_{j=1}^{n} a_{tj}|z_j|, \qquad t = 1, 2, \ldots, n. \tag{3}$$

In vector notation, (3) reads

$$|\lambda_i||z| \le A|z|.$$

Therefore, by Theorem 3.1(ii) and the definition of r,

$$|\lambda_i| \le f_A(|z|) \le r. \quad \blacksquare$$

Example 4.1. Let

$$A = \begin{bmatrix} 1 & 2 & 3 \\ 0 & 1 & 1 \\ 2 & 1 & 3 \end{bmatrix}.$$

Let $x = (0, 1, 1)$, $y = (I_3 + A)^2 x$, and $z = (I_3 + A)^2 y$. Compute $f_A(x)$, $f_A(y)$, and $f_A(z)$. Compare these numbers with r, the maximal eigenvalue of A.

We compute

$$\begin{aligned} f_A(x) &= \min(2, 4) = 2, \\ (I_3 + A)^2 &= \begin{bmatrix} 10 & 11 & 20 \\ 2 & 5 & 6 \\ 12 & 10 & 23 \end{bmatrix}, \\ y &= (31, 11, 33) \\ &= 75(0.4133\ldots, 0.1466\ldots, 0.44), \\ f_A(y) &= \min\left(\tfrac{152}{31}, \tfrac{44}{11}, \tfrac{172}{33}\right) = 4, \\ z &= (1091, 315, 1241) \\ &= 2647(0.4121\ldots, 0.1190\ldots, 0.4688\ldots), \\ f_A(z) &= \min\left(\tfrac{5444}{1091}, \tfrac{1556}{315}, \tfrac{6220}{1241}\right) \\ &= \tfrac{1556}{315} \\ &= 4.93968\ldots. \end{aligned}$$

We find by straightforward computation that $r = 5$, and the corresponding eigenvector in E^n is $(0.4117\ldots, 0.1176\ldots, 0.4705\ldots)$. $\blacksquare$

For nonnegative matrices that are not necessarily irreducible we can derive, by a continuity argument, the following weaker version of Theorem 4.1.

Theorem 4.2. *If A is a nonnegative $n \times n$ matrix, then A has a nonnegative eigenvalue r that is at least as large as the absolute value of any eigenvalue of A, and a nonnegative eigenvector corresponding to r.*

Proof. Let $A_\varepsilon = A + \varepsilon B$, where $\varepsilon > 0$ and B is any positive $n \times n$ matrix. Then $A_\varepsilon > 0$ and therefore, by Theorem 3.1, A_ε has a maximal eigenvalue r_ε such that

$$r_\varepsilon \geq |\lambda_i^{(\varepsilon)}|, \tag{4}$$

for any other eigenvalue $\lambda_i^{(\varepsilon)}$ of A_ε. (In fact, as we shall see later, $r_\varepsilon > |\lambda_i^{(\varepsilon)}|$ since $A_\varepsilon > 0$.) Also, there exists a positive vector $x^{(\varepsilon)}$ in E^n such that

$$A_\varepsilon x^{(\varepsilon)} = r_\varepsilon x^{(\varepsilon)}. \tag{5}$$

Now, eigenvalues and eigenvectors are continuous functions of entries of the matrix. Hence $A_\varepsilon \to A$ and $r_\varepsilon \to r$ as $\varepsilon \to 0$. Moreover, $\lambda_i^{(\varepsilon)} \to \lambda_i$, where the λ_i are eigenvalues of A, and, by (4),

$$r \geq |\lambda_i| \geq 0.$$

Also, (5) implies that

$$Ax = rx,$$

where $x \in E^n$ and therefore $x \neq 0$ is an eigenvector. ■

In case A is irreducible, we can refine the result in Theorem 4.1.

Theorem 4.3. *The maximal eigenvalue of an irreducible nonnegative matrix is a simple root of its characteristic equation.*

Proof. Let r be the maximal eigenvalue of A, an irreducible nonnegative $n \times n$ matrix, and let x be a nonnegative eigenvector corresponding to r. By Theorem 2.2, $x > 0$.

First, we show that the eigenspace of A corresponding to r is one-dimensional. Suppose that

$$Ay = ry,$$

where $y \neq 0$. Then, by the triangle inequality,

$$A|y| \geq r|y|, \tag{6}$$

where $|y|$ is a nonnegative nonzero vector.

Thus we conclude as in the proof of Theorem 4.1 that (6) is an equality and $|y|$ is a positive eigenvector of A corresponding to r. We have actually shown that an eigenvector of A corresponding to r cannot have zero entries. Next, suppose that $x = (x_1, x_2, \ldots, x_n)$ and $y = (y_1, y_2, \ldots, y_n)$ are nonzero vectors in the eigenspace of A corresponding to r. Then $|x| > 0$ and $|y| > 0$. Now, the vector $y_1x - x_1y$ is in the eigenspace of r but since its first coordinate is zero it cannot be an eigenvector. Hence

$$y_1x - x_1y = 0,$$

and x and y are linearly dependent. It follows that the eigenspace of r is of dimension 1.

We are now ready to prove that r is a simple root of the characteristic equation of A. Let $\Delta(\lambda) = \det(\lambda I_n - A)$. We show that $\Delta'(r) \neq 0$.

Recall that if every entry of $X = (x_{ij})$ is a differentiable function of λ, then

$$\frac{d}{d\lambda}(\det(X)) = \sum_{i,j=1}^{n} (-1)^{i+j}\det(X(i|j))\frac{d}{d\lambda}x_{ij}.$$

Thus

$$\begin{aligned}\Delta'(\lambda) &= \frac{d}{d\lambda}(\det(\lambda I_n - A)) \\ &= \sum_{i=1}^{n} \det((\lambda I_n - A)(i|i)) \\ &= \operatorname{tr}(\operatorname{adj}(\lambda I_n - A)),\end{aligned}$$

since

$$\frac{d}{d\lambda}((\lambda I_n - A)_{ij}) = \delta_{ij}.$$

Hence

$$\Delta'(r) = \operatorname{tr}(\operatorname{adj}(rI_n - A)).$$

Let $B(r) = \operatorname{adj}(rI_n - A)$. Then

$$\begin{aligned}(rI_n - A)B(r) &= I_n\det(rI_n - A) \\ &= 0. \end{aligned} \tag{7}$$

Since r has a one-dimensional eigenspace, the rank of $rI_n - A$ is $n - 1$ and therefore $B(r) \neq 0$. Suppose that the column $B(r)^{(j)}$ is different from zero. Then, by (7),

$$(rI_n - A)B(r)^{(j)} = 0,$$

that is, $B(r)^{(j)}$ is an eigenvector corresponding to r and thus $B(r)^{(j)}$ is a real multiple of a positive vector, namely, either $B(r)^{(j)} > 0$ or $B(r)^{(j)} < 0$. In other words, every column of $B(r)$ is either all positive or all negative or zero, and at least one of the columns is nonzero. Now, $(B(r))^{\mathrm{T}} = \operatorname{adj}(rI_n - A^{\mathrm{T}})$, and A^{T} is irreducible with maximal eigenvalue r. Hence the above conclusion applies also to columns of $(B(r))^{\mathrm{T}}$, that is, to rows of $B(r)$. Thus each row and each column of $B(r)$ is either positive or negative or zero, and at least one of the rows and one of the columns is nonzero. It follows that either

$$B(r) > 0,$$

or

$$B(r) < 0.$$

Thus

$$\Delta'(r) = \operatorname{tr}(B(r)) \neq 0,$$

and therefore r is a simple root of the characteristic equation of A. ■

Corollary 4.1. *Let A be an irreducible $n \times n$ matrix with maximal eigenvalue r and let $B(\lambda) = \operatorname{adj}(\lambda I_n - A)$. Then*

$$B(r) > 0.$$

The proof of Corollary 4.1 is quite easy, and we leave it as an exercise (Problem 11).

Corollary 4.2. *If A is an irreducible matrix with maximal eigenvalue r, and $Ax = rx$, then x is a scalar multiple of a positive vector.*

We have shown that the maximal eigenvalue of an irreducible matrix is simple. Nevertheless, the matrix may well have other positive eigenvalues. It is therefore remarkable that the maximal eigenvector of an irreducible matrix is (apart from positive multiples) the only nonnegative eigenvector the matrix can have.

Theorem 4.4. *An irreducible matrix has exactly one eigenvector in E^n.*

Proof. Let A be an irreducible matrix with maximal eigenvalue r, and let $y \in E^n$ be a maximal eigenvector of A^{T}. Then $y > 0$. Also, let $z \in E^n$ be any eigenvector of A,

$$Az = \zeta z.$$

If (,) denotes the standard inner product, then

$$\begin{aligned} \zeta(z, y) &= (Az, y) \\ &= (z, A^{\mathrm{T}}y) \\ &= r(z, y). \end{aligned}$$

Now, $(z, y) > 0$ since $z \in E^n$ and $y > 0$. Therefore $\zeta = r$. In other words, the only eigenvalue of A with nonnegative eigenvector is r. The result follows by Theorem 4.3. ■

We have shown that

$$r = \max\{ f_A(x) | x \in E^n \}$$
$$= \max\left\{ \min_{x_i \neq 0} \frac{(Ax)_i}{x_i} \middle| x \in E^n \right\}$$

is the maximal eigenvalue of A, and that r is a simple eigenvalue and has a positive eigenvector. Alternatively, we can arrive at the same conclusions by defining a function g_A:

$$g_A(x) = \max_{x_i \neq 0} \frac{(Ax)_i}{x_i}, \tag{8}$$

for $x \in \mathbb{P}^n$, $x \neq 0$, where we define $g_A(x) = +\infty$ if for some i, $x_i = 0$ and $(Ax)_i \neq 0$ [or equivalently we say that $g_A(x)$ is not defined for such x]. We show that

$$s = \min\{ g_A(x) | x \in E^n \}$$
$$= \min\left\{ \max_{x_i \neq 0} \frac{(Ax)_i}{x_i} \middle| x \in E^n \right\}$$

is the maximal eigenvalue of A.

Theorem 4.5. *Let A be an irreducible matrix and let g_A be the function defined in* (8). *Then*

(i) *g_A is homogeneous of degree* 0;
(ii) *if x is a nonnegative n-tuple such that $x_i > 0$ whenever $(Ax)_i > 0$, and σ is the least number for which*

$$\sigma x - Ax \geq 0,$$

then $\sigma = g_A(x)$;
(iii) *if x is as in* (ii) *and $y = (I_n + A)^{n-1}x$, then $g_A(y) \leq g_A(x)$;*
(iv) *g_A attains its minimum in E^n at a positive n-tuple.*

The proof of Theorem 4.5 is similar to those of Theorems 3.1 and 3.2. We leave it as an exercise (Problem 10).

Theorem 4.6. *Let A be an irreducible nonnegative $n \times n$ matrix with maximal eigenvalue r. If*

$$s = \min\{ g_A(x) | x \in E^n \},$$

then $s = r$.

Proof. Let x^0 be a positive n-tuple in E^n such that

$$g_A(x^0) \leq g_A(x),$$

for all $x \in E^n$ for which g_A is defined. We show that s is an eigenvalue and x^0 is an eigenvector of A. Since

$$sx^0 - Ax^0 \geq 0,$$

it suffices to prove that $sx^0 - Ax^0$ cannot be nonzero. Suppose that it is not zero. Then, by Corollary 2.1,

$$(I_n + A)^{n-1}(sx^0 - Ax^0) > 0,$$

that is,

$$sy^0 - Ay^0 > 0,$$

where $y^0 = (I_n + A)^{n-1}x^0$. Thus, for sufficiently small positive ε,

$$(s - \varepsilon)y^0 - Ay^0 \geq 0,$$

and therefore, by Theorem 4.5(ii),

$$g_A(y^0) \leq s - \varepsilon.$$

But this would imply that

$$g_A(y^0) < s,$$

which would contradict the minimality of s. Therefore

$$Ax^0 = sx^0,$$

and s is an eigenvalue of A. It follows from Theorem 4.4 that $s = r$. ■

We conclude this section with an important criterion for the irreducibility of a nonnegative matrix.

Theorem 4.7. *A matrix $A \geq 0$ with a simple maximal eigenvalue r is irreducible if and only if both A and A^{T} have positive eigenvectors corresponding to r.*

Proof. The necessity of the condition follows from Theorem 4.1. To prove the converse assume that $A \geq 0$ has a simple maximal eigenvalue r, and that both A and A^{T} have positive eigenvectors corresponding to r. If A were reducible, then there would exist a permutation matrix P such that

$$P^{\mathrm{T}}AP = \begin{bmatrix} B & D \\ 0 & C \end{bmatrix},$$

where B is a $k \times k$ submatrix. Note that $P^{\mathrm{T}}AP$ has a simple maximal eigenvalue r and both $P^{\mathrm{T}}AP$ and $P^{\mathrm{T}}A^{\mathrm{T}}P$ have positive eigenvectors corresponding to r. Now, r would have to be the maximal eigenvalue of either B or C. If $Bv = rv$, where v is a nonnegative nonzero k-tuple, then the n-tuple $v \dotplus 0$, which clearly is not a multiple of the positive eigenvector, would be an eigenvector of $P^{\mathrm{T}}AP$ corresponding to r. This would contradict the simplicity of r. If r were the maximal eigenvalue of C, then $C^{\mathrm{T}}u = ru$ for some nonnegative nonzero $(n-k)$-tuple u and $0 \dotplus u$ would be an eigenvector of $P^{\mathrm{T}}A^{\mathrm{T}}P$ linearly independent of the positive eigenvector. This again would contradict the simplicity of r. ■

1.5. PRINCIPAL SUBMATRICES OF NONNEGATIVE MATRICES

Submatrices of nonnegative matrices are, of course, nonnegative. The following results due to Frobenius [4] give a striking relation between the maximal eigenvalues of a nonnegative matrix and its principal submatrices.

Theorem 5.1. *The maximal eigenvalue of an irreducible matrix is greater than the maximal eigenvalue of any of its principal submatrices.*

Proof (Marcus and Minc [6]). Let A be an irreducible n-square matrix. We can assume without loss of generality that the principal submatrix in question lies in the first t rows and first t columns, that is,

$$A = \begin{bmatrix} B & C \\ D & E \end{bmatrix},$$

where B is the principal $t \times t$ submatrix. Let r and k be the maximal roots of A and B, respectively. Let y be a nonnegative eigenvector of B corresponding to k, and let $x = y \dotplus 0$ be the n-tuple whose first t coordinates are those of y, and whose last $n - t$ coordinates are 0. Then

$$\begin{aligned} Ax &= \begin{bmatrix} By \\ Dy \end{bmatrix} \\ &= k\begin{bmatrix} y \\ 0 \end{bmatrix} + \begin{bmatrix} 0 \\ Dy \end{bmatrix}, \end{aligned}$$

and therefore

$$\begin{aligned} Ax - kx &= \begin{bmatrix} 0 \\ Dy \end{bmatrix} \\ &\geq 0. \end{aligned}$$

Hence, by Theorem 3.1(ii),

$$k \leq f_A(x) < r.$$

The last inequality is strict by virtue of the fact that x is not a positive vector. ■

As in the proof of Theorem 4.2 we can use a continuity argument to obtain a similar though weaker result for nonnegative matrices that are not necessarily irreducible.

Theorem 5.2. *The maximal eigenvalue of a principal submatrix of a nonnegative matrix A cannot exceed the maximal eigenvalue of A.*

Of course, it is quite possible for the maximal eigenvalue of a principal submatrix of a nonnegative matrix A to be equal to the maximal eigenvalue of A. In fact, we have the following criterion for reducibility.

Theorem 5.3. *A nonnegative matrix A with maximal eigenvalue r is reducible if and only if r is an eigenvalue of a principal submatrix of A.*

Proof. Sufficiency follows immediately from Theorem 5.1. To prove the converse suppose that A is reducible, that is, A is cogredient to a matrix of the form

$$\begin{bmatrix} B & C \\ 0 & D \end{bmatrix},$$

where B and D are square. Then the spectrum of A consists of the eigenvalues of B together with those of D. Hence r must be either an eigenvalue of the principal submatrix B or of the principal submatrix D. ■

PROBLEMS

1 Show that a nonnegative $n \times n$ matrix A is reducible if and only if there exists a proper subset $\{e_{j_1}, e_{j_2}, \ldots, e_{j_k}\}$ of the standard basis of R^n such that

$$\langle Ae_{j_1}, Ae_{j_2}, \ldots, Ae_{j_k} \rangle \subset \langle e_{j_1}, e_{j_2}, \ldots, e_{j_k} \rangle.$$

2 Show that the $n \times n$ permutation matrix with 1's in positions $(1,2), (2,3), \ldots, (n-1, n)$, and $(n, 1)$, is irreducible.

3 Prove or disprove the following statements. (A and B denote arbitrary nonnegative $n \times n$ matrices.)

(a) If A is irreducible, then A^{T} is irreducible.

(b) If A is irreducible and p is an integer, then A^p is irreducible.

(c) If A^p is irreducible, then A is irreducible.

(d) If A and B are irreducible, then $A + B$ is irreducible.

(e) If A and B are irreducible, then AB is irreducible.

(f) If all eigenvalues of A are 0 then A is reducible.

(g) The matrix

$$\begin{bmatrix} 0 & 1 & 1 \\ 1 & 0 & 0 \\ 1 & 0 & 0 \end{bmatrix}$$

is reducible.

(h) The matrix

$$\begin{bmatrix} 0 & 0 & 1 & 0 \\ 0 & 0 & 0 & 1 \\ 0 & 1 & 0 & 0 \\ 1 & 0 & 0 & 0 \end{bmatrix}$$

is reducible.

(i) The matrix

$$\begin{bmatrix} 0 & 0 & 1 & 0 \\ 0 & 0 & 1 & 1 \\ 1 & 0 & 0 & 0 \\ 1 & 1 & 0 & 0 \end{bmatrix}$$

is reducible.

(j) A is irreducible if and only if $I_n + A$ is irreducible.

(k) If AB is irreducible, then BA is irreducible.

4 Prove that if A is an irreducible matrix with minimal polynomial of degree m, then for every (i, j), $i \neq j$, there exists a positive integer k not exceeding $m - 1$ such that $a_{ij}^{(k)} > 0$.

5 Prove that if A is a nonnegative matrix and $A^2 = I_n$, then either $A = I_n$, or A is cogredient to a direct sum of 2×2 matrices of the form

$$\begin{bmatrix} 0 & a^{-1} \\ a & 0 \end{bmatrix}$$

and, possibly, an identity matrix [5].

6 Show that a permutation matrix is irreducible if and only if it is nonderogatory.

7 Let A be an irreducible nonnegative $n \times n$ matrix with a positive trace. Show that $A^k > 0$ for a sufficiently large integer k.

8 Let A be a nonnegative $n \times n$ matrix. Show that if $(I_n + A)x$ has less zero entries than x, for every nonnegative nonzero n-tuple x with some zero entries, then A is irreducible.

9 Let $f(x_1, x_2, \ldots, x_m)$ be a polynomial in indeterminates $x_1, x_2, \ldots, x_m$ with complex coefficients. Show that if $f(a_1, a_2, \ldots, a_m) = 0$ for all nonnegative $a_1, a_2, \ldots, a_m$, then $f(x_1, x_2, \ldots, x_m) = 0$. (*Hint*: Use induction on m and express $f(x_1, x_2, \ldots, x_m)$ as a polynomial in x_m with coefficients in $\mathbb{C}[x_1, x_2, \ldots, x_{m-1}]$.)

10 Prove Theorem 4.5.

11 Prove Corollary 4.1.

12 Let $A \in M_n(\mathbb{P})$ be irreducible and let f_A be the Collatz–Wielandt function associated with A. Find a necessary and sufficient condition on A that f_A be continuous on all of E^n.

13 Let G be the set defined in the proof of Theorem 3.2. Find the maximum and the minimum of $\sum_{i=1}^{n} y_i$ as $(y_1, y_2, \ldots, y_n)$ runs over all n-tuples in G.

14 Let

$$A = \begin{bmatrix} 1 & 3 & 4 \\ 2 & 1 & 1 \\ 3 & 4 & 5 \end{bmatrix},$$

and let f_A be the Collatz–Wielandt function and g_A the function defined in Section 1.4. Compute $f_A(x)$ and $g_A(x)$ for each of the following x: $(1, 1, 1)$, $(1, 0, 2)$, and $(2, 1, 3)$. Deduce a lower bound and an upper bound for the maximal eigenvalue of A.

15 Let A be the matrix in Problem 14. Let $y = (I_3 + A)^2 x$ where $x = (2, 1, 3)$. Compute $f_A(y)$ and $g_A(y)$. Use these data to obtain a lower bound and an upper bound for the maximal eigenvalue of A. How do these bounds compare with those in Problem 14?

16 Let A and B be nonnegative $n \times n$ matrices. Show that if A is irreducible and $B \neq 0$, then the maximal eigenvalue of $A + B$ is greater than that of A.

17 Let C be a principal submatrix of a nonnegative matrix A with maximal eigenvalue r. If r is an eigenvalue of C show that r is also an eigenvalue of every principal submatrix of A containing C.

18 Let $B \geq 0$, $x \in E^n$, and let $y = (y_1, y_2, \ldots, y_n) = Bx$. Show that

$$\min_j c_j \leq \sum_{i=1}^{n} y_i \leq \max_j c_j,$$

where the c_j are the column sums of B, $c_j = \sum_{t=1}^{n} b_{tj}$, $j = 1, 2, \ldots, n$.

REFERENCES

1. L. Collatz, Einschliessungssatz für die charakteristischen Zahlen von Matrizen, *Math. Z.* **48** (1942), 221–226.

2. G. Frobenius, Über die Darstellung der endlichen Gruppen durch lineare Substitutionen, *S.-B. K. Preuss. Akad. Wiss. Berlin* (1897), 994–1015.

3. G. Frobenius, Über Matrizen aus positiven Elementen, *S.-B. K. Preuss. Akad. Wiss. Berlin* (1908), 471–476; (1909), 514–518.

4. G. Frobenius, Über Matrizen aus nicht negativen Elementen, *S.-B. K. Preuss. Akad. Wiss. Berlin* (1912), 456–477.

5. F. Harary and H. Minc, Which nonnegative matrices are self-inverse? *Math. Mag.* **49** (1976), 91–92.

6. M. Marcus and H. Minc, Two theorems of Frobenius, *Pacific J. Math.* **60** (1975), 149–151.

7. H. Minc, Linear transformations on nonnegative matrices, *Linear Algebra Appl.* **9** (1974), 149–153.

8. O. Perron, Zur Theorie der Matrizen, *Math. Ann.* **64** (1907), 248–263.

9. H. Wielandt, Unzerlegbare nicht-negative Matrizen, *Math. Z.* **52** (1950), 642–648.

II

Localization of the Maximal Eigenvalue

2.1. BOUNDS FOR THE MAXIMAL EIGENVALUE OF A NONNEGATIVE MATRIX

The problem of localizing the maximal eigenvalue of a nonnegative matrix is of importance not only in theoretical mathematics but also in computations where iterative processes require an initial estimate of the maximal eigenvalue. Such estimates are particularly useful if the bounds are expressed in terms of easily computable functions of the entries of the matrix, such as row sums or column sums.

The best known and most frequently used bounds for the maximal eigenvalue of a nonnegative matrix are due to Frobenius [2]. Let r_i and c_j denote the ith row sum and the jth column sum of a matrix $A = (a_{ij}) \in M_n(\mathbb{C})$, respectively, that is,

$$r_i = \sum_{t=1}^{n} a_{it}, \qquad i = 1, 2, \ldots, n,$$

and

$$c_j = \sum_{t=1}^{n} a_{tj}, \qquad j = 1, 2, \ldots, n.$$

Theorem 1.1. *If A is a nonnegative matrix with maximal eigenvalue r and row sums $r_1, r_2, \ldots, r_n$, then*

$$\rho \leq r \leq R, \tag{1}$$

where $\rho = \min_i r_i$, and $R = \max_i r_i$. If A is irreducible, then equality can hold on either side of (1) *if and only if all row sums of A are equal.*

Clearly, an analogous result holds for column sums.

Proof. We first assume that A is irreducible. Let $(x_1, x_2, \ldots, x_n) > 0$ be a maximal eigenvector. Then

$$\sum_{j=1}^{n} a_{ij}x_j = rx_i, \qquad i = 1, 2, \ldots, n.$$

Now, if $x_m \geq x_j$ for $j = 1, 2, \ldots, n$, then

$$\begin{aligned} r &= \frac{1}{x_m}\sum_{j=1}^{n} a_{mj}x_j \\ &\leq \sum_{j=1}^{n} a_{mj} \\ &= r_m \\ &\leq R. \end{aligned}$$

Similarly, if x_μ is the least of coordinates x_j, then

$$\begin{aligned} r &= \frac{1}{x_\mu}\sum_{j=1}^{n} a_{\mu j}x_j \\ &\geq \sum_{j=1}^{n} a_{\mu j} \\ &= r_\mu \\ &\geq \rho. \end{aligned}$$

If A happens to be reducible, then the result follows by a continuity argument, in the same manner as Theorems 4.2 and 5.2 in Chapter I follow from Theorems 4.1 and 5.1, respectively. ■

An alternative proof, which easily yields the conditions for equality, is an immediate consequence of the following elementary lemma.

Lemma 1.1. *Let α be an eigenvalue of A and let $(x_1, x_2, \ldots, x_n)$ and $(y_1, y_2, \ldots, y_n)$ be eigenvectors corresponding to α of A^{T} and A, respectively. Then*

$$\alpha\sum_{i=1}^{n} x_i = \sum_{t=1}^{n} x_t r_t, \tag{2}$$

and

$$\alpha\sum_{j=1}^{n} y_j = \sum_{t=1}^{n} y_t c_t. \tag{3}$$

Proof. We have

$$\alpha x_i = \sum_{t=1}^{n} a_{ti}x_t, \qquad i = 1, 2, \ldots, n.$$

We sum both sides with respect to i,

$$\alpha \sum_{i=1}^{n} x_i = \sum_{i=1}^{n} \sum_{t=1}^{n} a_{ti} x_t$$
$$= \sum_{t=1}^{n} x_t \sum_{i=1}^{n} a_{ti}$$
$$= \sum_{t=1}^{n} x_t r_t .$$

Formula (3) is proved similarly. ■

Now, let A be a nonnegative matrix with maximal eigenvalue r. Let $(x_1, x_2, \ldots, x_n) \geq 0$ be a maximal eigenvector of A^{T}, and let $\sum_{i=1}^{n} x_i = 1$. Then, by (2),

$$r = \sum_{t=1}^{n} x_t r_t, \tag{4}$$

and therefore

$$\min_t r_t \leq r \leq \max_t r_t . \tag{5}$$

Moreover, if A is irreducible, then $(x_1, x_2, \ldots, x_n) > 0$. In this case equality on either side of (5) can hold if and only if

$$r = r_t,$$

for $t = 1, 2, \ldots, n$. In other words, if A is the matrix described in the statement of Theorem 1.1, $\rho < R$, and A is irreducible, then

$$\rho < r < R. \quad ■ \tag{6}$$

Theorem 1.1 can also be proved for an irreducible matrix A by considering $f_A(u)$ and $g_A(u)$ for an appropriate vector u (see Problem 2). The result can then be extended to all nonnegative matrices by a continuity argument.

Equality (4) can be used to improve Frobenius' bounds. We shall require the following inequality:

If $q_1, q_2, \ldots, q_n$ are positive numbers, then

$$\min_i \frac{p_i}{q_i} \leq \frac{p_1 + p_2 + \cdots + p_n}{q_1 + q_2 + \cdots + q_n} \leq \max_i \frac{p_i}{q_i}, \tag{7}$$

for any real numbers $p_1, p_2, \ldots, p_n$. Equality holds on either side of (7) *if and only if all the ratios p_i/q_i are equal.* The proof of (7) is elementary (see, e.g., [4]).

Theorem 1.2. *Let $A = (a_{ij})$ be a nonnegative matrix with nonzero row sums $r_1, r_2, \ldots, r_n$ and maximal eigenvalue r. Then*

$$\min_i \left(\frac{1}{r_i} \sum_{t=1}^{n} a_{it} r_t \right) \le r \le \max_i \left(\frac{1}{r_i} \sum_{t=1}^{n} a_{it} r_t \right). \tag{8}$$

Proof. We assume that A is irreducible. The result can be then extended to reducible matrices by a continuity argument. Let $r_i(A^2)$ denote the ith row sum of matrix A^2. Applying formula (4) to matrices A^2 and A we have

$$r^2 = \sum_{i=1}^{n} x_i r_i(A^2),$$

and

$$r = \sum_{i=1}^{n} x_i r_i,$$

where $(x_1, x_2, \ldots, x_n)$ is the maximal eigenvector of A^T [and thus of $(A^2)^T$] with $\sum_{i=1}^{n} x_i = 1$. Therefore

$$r = \frac{r^2}{r} = \frac{\sum_{i=1}^{n} x_i r_i(A^2)}{\sum_{i=1}^{n} x_i r_i},$$

and, by (7),

$$\min_i \frac{r_i(A^2)}{r_i} \le \min_{x_i \ne 0} \frac{r_i(A^2)}{r_i} \le r \le \max_{x_i \ne 0} \frac{r_i(A^2)}{r_i} \le \max_i \frac{r_i(A^2)}{r_i}.$$

Now,

$$\begin{aligned} r_i(A^2) &= \sum_{j=1}^{n} \sum_{t=1}^{n} a_{it} a_{tj} \\ &= \sum_{t=1}^{n} a_{it} \sum_{j=1}^{n} a_{tj} \\ &= \sum_{t=1}^{n} a_{it} r_t, \end{aligned}$$

and the result follows. ■

Alternatively, we can show that (8) follows from Theorem 1.1. Let

$$D = \operatorname{diag}(r_1, r_2, \ldots, r_n).$$

Note that the ith row sum of $D^{-1}AD$ is

$$\frac{1}{r_i}\sum_{t=1}^{n} a_{it}r_t.$$

Now, apply Theorem 1.1 to the matrix $D^{-1}AD$ (which is, of course, similar to A):

$$\min_i\left(\frac{1}{r_i}\sum_{t=1}^{n} a_{it}r_t\right) \le r \le \max_i\left(\frac{1}{r_i}\sum_{t=1}^{n} a_{it}r_t\right). \quad \blacksquare$$

We also observe that the bounds in (8) are sharper than those in (1). For, by (7),

$$\min_t r_t \le \frac{\sum_{t=1}^{n} a_{it}r_t}{\sum_{t=1}^{n} a_{it}\cdot 1} \le \max_t r_t,$$

that is,

$$\rho = \min_t r_t \le \frac{1}{r_i}\sum_{t=1}^{n} a_{it}r_t \le \max_t r_t = R,$$

for $i = 1, 2, \ldots, n$.

If A is a positive matrix with maximal eigenvalue r, maximum row sum R, and minimum row sum ρ, and if $\rho < R$, then as noted in (6),

$$\rho < r < R.$$

Ledermann proposed the problem of determining positive numbers p_1 and p_2 such that

$$\rho + p_1 \le r \le R - p_2.$$

He obtained the following result.

Theorem 1.3 (Ledermann [3]). *Let $A = (a_{ij})$ be a positive matrix with maximal eigenvalue r and row sums $r_1, r_2, \ldots, r_n$. If $R = \max_i r_i$, $\rho = \min_i r_i$, $\eta = \min_{i,j} a_{ij}$, and $R > \rho$, then*

$$\rho + \eta\left(\frac{1}{\sqrt{\delta}} - 1\right) \le r \le R - \eta(1 - \sqrt{\delta}), \tag{9}$$

where $\delta = \max_{r_i < r_j}(r_i/r_j)$.

Proof. Let $(x_1, x_2, \ldots, x_n)$ be a positive maximal eigenvector of A. We can assume without loss of generality that $1 = x_1 \ge x_2 \ge \cdots \ge x_n > 0$.

[Such an ordering of the x_j can be achieved by premultiplying A by a suitable permutation matrix P and postmultiplying it by P^{-1} (see Problem 3).] Clearly, $r_n < r < r_1$ (see Problem 4). Therefore $r_n/r_1 < 1$, and thus $r_n/r_1 \leq \delta$. Now,

$$rx_n = \sum_{j=1}^{n} a_{nj}x_j < r_n,$$

and

$$r = \sum_{j=1}^{n} a_{1j}x_j > x_n r_1.$$

It follows that

$$x_n < \frac{r_n}{x_n r_1}.$$

Hence

$$x_n < \sqrt{\frac{r_n}{r_1}} \leq \sqrt{\delta}\,.$$

Thus

$$\begin{aligned} r &= \sum_{j=1}^{n} a_{1j}x_j \\ &< \sum_{j=1}^{n-1} a_{1j} + a_{1n}\sqrt{\delta} \\ &= r_1 - a_{1n}(1 - \sqrt{\delta}\,) \\ &\leq R - \eta(1 - \sqrt{\delta}\,). \end{aligned}$$

Similarly,

$$\begin{aligned} r &= \frac{1}{x_n}\sum_{j=1}^{n} a_{nj}x_j \\ &> \frac{a_{n1}}{\sqrt{\delta}} + \sum_{j=2}^{n} a_{nj} \\ &= r_n + a_{n1}\left(\frac{1}{\sqrt{\delta}} - 1\right) \\ &\geq \rho + \eta\left(\frac{1}{\sqrt{\delta}} - 1\right). \quad \blacksquare \end{aligned}$$

Ledermann's result was improved by Ostrowski.

Theorem 1.4 (Ostrowski [7]). *Let $A = (a_{ij})$ be a positive $n \times n$ matrix with maximal eigenvalue r and row sums $r_1, r_2, \dots, r_n$. Let $\sigma = \sqrt{(\rho - \eta)/(R - \eta)}$, where $R = \max_i r_i$, $\rho = \min_i r_i$, and $\eta = \min_{i,j} a_{ij}$. Then*

$$\rho + \eta\left(\frac{1}{\sigma} - 1\right) \le r \le R - \eta(1 - \sigma).$$

Proof. Let $(x_1, x_2, \dots, x_n)$ be a maximal eigenvector. For simplicity, assume that $1 = x_1 \ge x_2 \ge \cdots \ge x_n > 0$. Then

$$\begin{aligned} rx_i &= \sum_{j=1}^{n} a_{ij}x_j \\ &\ge a_{i1} + x_n \sum_{j=2}^{n} a_{ij} \\ &= a_{i1}(1 - x_n) + r_i x_n, \qquad i = 1, 2, \dots, n. \end{aligned}$$

Therefore

$$r \ge (x_n r_i + \eta(1 - x_n))/x_i, \qquad i = 1, 2, \dots, n. \tag{10}$$

Similarly,

$$\begin{aligned} rx_i &\le \sum_{j=1}^{n-1} a_{ij} + a_{in}x_n \\ &= r_i - a_{in}(1 - x_n), \end{aligned}$$

and therefore

$$r \le (r_i - \eta(1 - x_n))/x_i, \qquad i = 1, 2, \dots, n. \tag{11}$$

Let $r_s = R$ and $r_t = \rho$. Setting $i = s$ in (10), we have

$$\begin{aligned} r &\ge (x_n R + \eta(1 - x_n))/x_s \\ &\ge x_n R + \eta(1 - x_n) \\ &= x_n(R - \eta) + \eta. \end{aligned} \tag{12}$$

Similarly, substituting t for i in (11) we obtain

$$r \le \frac{\rho - \eta}{x_n} + \eta. \tag{13}$$

Hence, from (12) and (13),

$$x_n(R - \eta) \le r - \eta \le \frac{\rho - \eta}{x_n},$$

and therefore

$$x_n \leq \sqrt{\frac{\rho - \eta}{R - \eta}} = \sigma.$$

Now, put $i = n$ in (10) and $i = 1$ in (11),

$$r_n + \eta\left(\frac{1}{x_n} - 1\right) \leq r \leq r_1 - \eta(1 - x_n).$$

Thus a fortiori

$$\rho + \eta\left(\frac{1}{\sigma} - 1\right) \leq r \leq R - \eta(1 - \sigma). \quad \blacksquare$$

Note that Ostrowski's bounds are sharper than those in Theorem 1.3:

$$\sigma^2 = \frac{\rho - \eta}{R - \eta} \leq \frac{\rho}{R} \leq \max_{r_i < r_j} \frac{r_i}{r_j} = \delta$$

(see Example 1.1 below).

Brauer [1] improved these bounds for the maximal eigenvalue of a positive matrix and showed that his is the best possible result involving η, R, and ρ, in the sense that for any prescribed R, ρ, and η satisfying $R > \rho \geq n\eta > 0$, there exist positive $n \times n$ matrices with maximum row sum R, minimum row sum ρ, and minimum entry η, for which Brauer's bounds are attained.

Theorem 1.5 (Brauer [1]). *Let A, r, R, ρ, and η be as defined in the statement of Theorem 1.4. Let*

$$g = \frac{R - 2\eta + \sqrt{R^2 - 4\eta(R - \rho)}}{2(\rho - \eta)}, \qquad h = \frac{-\rho + 2\eta + \sqrt{\rho^2 + 4\eta(R - \rho)}}{2\eta}.$$

Then

$$\rho + \eta(h - 1) \leq r \leq R - \eta(1 - 1/g).$$

Proof. The method of proof is to apply the bounds in Theorem 1.1 to two matrices obtained from A by means of suitable similarity transformations.

Let $r_1, r_2, \ldots, r_n$ denote the row sums of A. We can assume without loss of generality that $r_1 = R$ and $r_n = \rho$. Let B be the matrix obtained from A by multiplying the last row of A by g and its last column by $1/g$. Obviously, A and B are similar and have the same spectrum. The ith row sum of B, $i = 1, 2, \ldots, n - 1$, is

$$\begin{aligned} r_i - a_{in}(1 - 1/g) &\leq R - \eta(1 - 1/g) \\ &= K_1, \quad \text{say}. \end{aligned}$$

The nth row sum of B is equal to

$$\begin{aligned} g\rho - a_{nn}(g-1) &\le g\rho - \eta(g-1) \\ &= K_2, \quad \text{say.} \end{aligned}$$

A straightforward computation shows that if g is as defined in the statement of the theorem, then $K_1 = K_2$. Hence all row sums of B are bounded above by $R - \eta(1 - 1/g)$, and, by Theorem 1.1,

$$r \le R - \eta(1 - 1/g).$$

In order to obtain the lower bound, we construct matrix C obtained from A by multiplying its first row by $1/h$, and its first column by h. Then A and C are similar. The first row sum of C is

$$R/h + a_{11}(1 - 1/h) \ge R/h + \eta(1 - 1/h) = K_3,$$

and the ith row sum of C, $i = 2, 3, \ldots, n$, is

$$r_i + a_{i1}(h-1) \ge \rho + \eta(h-1) = K_4.$$

Again, a straightforward computation shows that if h has the prescribed value, then $K_3 = K_4$. Thus all row sums of C are bounded below by $\rho + \eta(h-1)$, and, by Theorem 1.1,

$$r \ge \rho + \eta(h-1). \quad \blacksquare$$

We now construct two matrices Q and Q' with prescribed R, ρ, and η, $R > \rho \ge n\eta > 0$, whose maximal eigenvalues attain the upper and lower bound in Theorem 1.5, respectively. Let

$$Q = \left[\begin{array}{c|c} P_1 & P_2 \\ \hline P_3 & \eta \end{array}\right],$$

where every entry is not less than η, all the row sums of the $(n-1) \times (n-1)$ submatrix P_1 are equal to $R - \eta$, every entry in P_2 is η, and all those in P_3 add up to $\rho - \eta$. Now, Q is similar to

$$\left[\begin{array}{c|c} P_1 & \frac{1}{g}P_2 \\ \hline gP_3 & \eta \end{array}\right],$$

each of whose first $n-1$ row sums is equal to $R - \eta + \eta/g = K_1$, and its last row sum is $g\rho - g\eta + \eta = K_2 = K_1$. Hence, by Theorem 1.1, its maximal eigenvalue and that of Q are equal to K_1.

Similarly, we construct a matrix

$$Q' = \left[\begin{array}{c|c} \eta & P_2' \\ \hline P_3' & P_4' \end{array}\right],$$

where no entry is less than η, every row sum of the $(n-1) \times (n-1)$ submatrix P_4' is equal to $\rho - \eta$, each entry in P_3' is η, and those of P_2' add up to $R - \eta$. Now, if the first row of Q' is multiplied by $1/h$, and its first column by h, then the resulting matrix is similar to Q', and all its row sums are $K_3 = K_4$.

Many other bounds for the maximal eigenvalue of a nonnegative matrix are known. However, most of them are rather complicated and of not much interest, at least in the context of this book.

Example 1.1. Compute the bounds given in Theorems 1.1 (Frobenius), 1.2, 1.3 (Ledermann), 1.4 (Ostrowski), and 1.5 (Brauer), for the maximal eigenvalue r of

$$A = \begin{bmatrix} 1 & 1 & 2 \\ 2 & 3 & 3 \\ 4 & 1 & 1 \end{bmatrix}.$$

Frobenius' bounds:	$4 < r < 8$ (rows),
	$5 < r < 7$ (columns);
Theorem 1.2:	$5 < r < 6.25$ (rows),
	$5.6 < r < 5.8572$ (columns);
Ledermann's bounds:	$4.1547 < r < 7.8661$ (rows),
	$5.080 < r < 6.9259$ (columns);
Ostrowski's bounds:	$4.5275 < r < 7.6547$ (rows),
	$5.2247 < r < 6.8165$ (columns);
Brauer's bounds:	$4.8284 < r < 7.4642$ (rows),
	$5.3722 < r < 6.7016$ (columns).

The maximal eigenvalue actually is $r = 5.74165738\ldots$. ■

Example 1.2. Find a matrix diagonally similar to the matrix A in Example 1.1 for which Theorem 1.1 produces sharper bounds for the maximal eigenvalue r of A than those obtained in Example 1.1.

We first try to find a matrix diagonally similar to A whose maximum column sum is less than 7, the maximum column sum of A. Let $X = \operatorname{diag}(x, 1, 1)$, where x is a positive number to be determined. The column sums of $X^{-1}AX$ are

$$1 + 2x + 4x, \qquad \frac{1}{x} + 3 + 1, \quad \text{and} \quad \frac{2}{x} + 3 + 1.$$

If $x < 1$, then $1 + 6x < 7$, as intended. However, the other two column sums increase as x decreases. Therefore the least value that the maximum row sum of $X^{-1}AX$ can have is given by

$$1 + 6x = \frac{2}{x} + 4,$$

that is,

$$x = \tfrac{1}{12}(3 + \sqrt{57}),$$

which yields the upper bound

$$r < \tfrac{1}{2}(5 + \sqrt{57}) < 6.2750.$$

Similarly, in order to improve the lower bound for r, we set $Y = \operatorname{diag}(1, y, 1)$. Then the column sums of $Y^{-1}AY$ are

$$1 + \frac{2}{y} + 4, \qquad y + 3 + y, \quad \text{and} \quad 2 + \frac{3}{y} + 1.$$

Our purpose is to find $y > 1$ for which

$$5 < 2y + 3 \leq \min\left(5 + \frac{2}{y}, 3 + \frac{3}{y}\right).$$

Clearly, the optimal choice is given by

$$2y + 3 = 3 + \frac{3}{y},$$

that is,

$$y = \sqrt{\tfrac{3}{2}},$$

which gives the lower bound

$$r > \sqrt{6} + 3 > 5.4494.$$

An analogous method applied to row sums of A yields the following bounds:

$$5 < r < 6.4495. \quad \blacksquare$$

Example 1.3. Find bounds for the maximal eigenvalue r of the matrix in Example 1.1, transforming the matrix twice by diagonal similarity transformations, as in the second proof of Theorem 1.2.

Let $D_1 = \text{diag}(4, 8, 6)$. Then

$$D_1^{-1}AD_1 = \begin{bmatrix} 1 & 2 & 3 \\ 1 & 3 & \frac{9}{4} \\ \frac{8}{3} & \frac{4}{3} & 1 \end{bmatrix}.$$

The row sums of this matrix are 6, $\frac{25}{4}$, and 5. Thus

$$D_2^{-1}D_1^{-1}AD_1D_2 = \begin{bmatrix} 1 & \frac{25}{12} & \frac{5}{2} \\ \frac{24}{25} & 3 & \frac{9}{5} \\ \frac{16}{5} & \frac{5}{3} & 1 \end{bmatrix},$$

where $D = \text{diag}(6, \frac{25}{4}, 5)$. Now, applying Theorem 1.1 to rows of the above matrix yields the bounds

$$5.5833 < r < 5.8667.$$

We can also apply the same method to columns. Let $G_1 = \text{diag}(7, 5, 6)$. Then

$$G_1AG_1^{-1} = \begin{bmatrix} 1 & \frac{7}{5} & \frac{7}{3} \\ \frac{10}{7} & 3 & \frac{5}{2} \\ \frac{24}{7} & \frac{6}{5} & 1 \end{bmatrix},$$

whose column sums are $\frac{41}{7}$, $\frac{28}{5}$, and $\frac{35}{6}$. If $G_2 = \text{diag}(\frac{41}{7}, \frac{28}{5}, \frac{35}{6})$, then

$$G_2G_1AG_1^{-1}G_2^{-1} = \begin{bmatrix} 1 & \frac{41}{28} & \frac{82}{35} \\ \frac{56}{41} & 3 & \frac{12}{5} \\ \frac{140}{41} & \frac{5}{4} & 1 \end{bmatrix}.$$

Applying Theorem 1.1 to columns of this matrix, we obtain

$$5.7142 < r < 5.7805. \quad \blacksquare$$

Example 1.4. Construct 3×3 matrices Q and Q' with $R = 8$, $\rho = 4$, $\eta = 1$, and $r = 7.464\ldots$ and $4.828\ldots$, respectively (see Example 1.1).

Let

$$Q = \begin{bmatrix} 4 & 3 & 1 \\ 4 & 3 & 1 \\ 2 & 1 & 1 \end{bmatrix}$$

(see the construction of Q after the proof of Theorem 1.5). The matrix

$$\begin{bmatrix} 4 & 3 & 1/g \\ 4 & 3 & 1/g \\ 2g & g & 1 \end{bmatrix},$$

where $g = 2.15470053\ldots$, is similar to Q, and all its row sums are equal to $7.46410161\ldots$, which is therefore the maximal eigenvalue of Q.

Similarly, let

$$Q' = \begin{bmatrix} 1 & 4 & 3 \\ 1 & 2 & 1 \\ 1 & 2 & 1 \end{bmatrix}.$$

Then the matrix

$$\begin{bmatrix} 1 & 4/h & 3/h \\ h & 2 & 1 \\ h & 2 & 1 \end{bmatrix},$$

where $h = 1.82842712\ldots$, is similar to Q', and all its row sums are $4.82842712\ldots$ which is therefore the maximal eigenvalue of Q'. ■

2.2. DOMINATING NONNEGATIVE MATRIX

If $C = (c_{ij})$ is an $n \times n$ complex matrix and $A = (a_{ij})$ is an $n \times n$ nonnegative matrix such that $|C| \le A$ (i.e., $|c_{ij}| \le a_{ij}$ for all i, j), then A is said to *dominate* C. The following remarkable result is due to Wielandt [9].

Theorem 2.1. *If a complex matrix C is dominated by an irreducible matrix A with maximal eigenvalue r, then for every eigenvalue s of C,*

$$|s| \le r. \tag{1}$$

Equality holds in (1) *if and only if*

$$C = e^{i\varphi} DAD^{-1}, \tag{2}$$

where $s = re^{i\varphi}$ and $|D| = I_n$.

Proof. Let

$$Cy = sy, \tag{3}$$

where $y \neq 0$. Then

$$|C|\,|y| \ge |s|\,|y|,$$

by the triangle inequality. But $A \geq |C|$, and therefore

$$A|y| \geq |C||y| \geq |s||y|. \tag{4}$$

Hence, by Theorem 3.1(ii), Chapter I,

$$|s| \leq f_A(|y|),$$

where f_A is the Collatz–Wielandt function associated with A, and therefore

$$|s| \leq f_A(|y|) \leq r. \tag{5}$$

Suppose that $C = e^{i\varphi}DAD^{-1}$, where $|D| = I_n$. Then the matrices C and $e^{i\varphi}A$ are similar and if r is the maximal eigenvalue of A, then $re^{i\varphi}$ is an eigenvalue of C.

We now prove the necessity of the condition (2) for equality in (1). Suppose that $s = re^{i\varphi}$, that is, $|s| = r$. Then (5) implies that $f_A(|y|) = r$. Thus $|y|$ is a maximal vector, and, by (4),

$$A|y| = |C||y| = r|y|, \tag{6}$$

that is,

$$(A - |C|)|y| = 0.$$

Since $|y|$ is maximal, we must have $|y| > 0$ (see Corollary 4.2, Chapter I). Also $A - |C| \geq 0$, and therefore

$$A = |C|. \tag{7}$$

Define

$$D = \operatorname{diag}\left(\frac{y_1}{|y_1|}, \frac{y_2}{|y_2|}, \ldots, \frac{y_n}{|y_n|}\right),$$

and

$$G = (g_{ij}) = e^{-i\varphi}D^{-1}CD.$$

Then (3) gives

$$CD|y| = sD|y| = re^{i\varphi}D|y|.$$

Therefore

$$G|y| = r|y|,$$

and, by (6),

$$G|y| = A|y|. \tag{8}$$

Now, by the definition of G,

$$|G| = |C|$$

and, by (7),

$$|G| = A.$$

Hence (8) yields

$$|G|\,|y| = G|y|,$$

or

$$\sum_{j=1}^{n} \left(|g_{ij}| - g_{ij}\right)|y_j| = 0, \qquad i = 1, 2, \ldots, n,$$

which implies that

$$|g_{ij}| - g_{ij} = 0,$$

for all i and j, since $|y_j| > 0$, $j = 1, 2, \ldots, n$. It follows that

$$G = |G| = A,$$

and thus from the definition of G,

$$C = e^{i\varphi} DAD^{-1}. \quad \blacksquare$$

Corollary 2.1. *Let A be a nonnegative matrix with maximal eigenvalue r, and let C be a complex matrix such that $|C| \leq A$. Then for any eigenvalue s of C,*

$$|s| \leq r.$$

The corollary follows easily from the theorem by a continuity argument. The condition for equality is not, of course, necessary if A is reducible.

Corollary 2.2. *If A is an irreducible matrix, and $A \geq C \geq 0$, $A \neq C$, then $r(A) > r(C)$.*

Example 2.1. Let A be an $n \times n$ nonnegative matrix with maximal eigenvalue r, and let ρ be the maximal eigenvalue of a principal submatrix of A. Use Theorem 2.1 to prove that

$$\rho \leq r,$$

and that the inequality is strict in case A is irreducible (see Theorem 5.1, Chapter I).

Let $A[i_1, i_2, \ldots, i_k | i_1, i_2, \ldots, i_k]$ be the principal submatrix, and C be the $n \times n$ matrix such that $C[i_1, i_2, \ldots, i_k | i_1, i_2, \ldots, i_k] = A[i_1, i_2, \ldots, i_k | i_1, i_2, \ldots, i_k]$ and all other entries of C are 0. Then $C \leq A$, and the result follows by Theorem 2.1. ■

Our next result gives bounds for the difference of maximal eigenvalues of two irreducible matrices one of which dominates the other.

Theorem 2.2 (Marcus, Minc, and Moyls [5]). *Let $A = (a_{ij})$ and $B = (b_{ij})$ be irreducible matrices with maximum eigenvalues α and β, respectively, and let $A \leq B$. Let $S = (s_{ij})$ be any nonnegative matrix which commutes either with A or with B and has nonzero column sums $c_1, c_2, \ldots, c_n$. Then*

$$\frac{m}{\max_{i,j}(s_{ij}/c_j)} \leq \beta - \alpha \leq \frac{M}{\min_{i,j}(s_{ij}/c_j)},$$

where $m = \min_{i,j}(b_{ij} - a_{ij})$ and $M = \max_{i,j}(b_{ij} - a_{ij})$.

Proof. Let z be the maximal eigenvector of A, $z \in E^n$, and let f_B be the Collatz–Wielandt function associated with B. Suppose that t is the integer for which

$$f_B(z) = \frac{(Bz)_t}{z_t}.$$

Since

$$(B - A)z + Az = Bz,$$

that is,

$$(B - A)z + \alpha z = Bz,$$

then, in particular,

$$\sum_{j=1}^{n} (b_{tj} - a_{tj})z_j + \alpha z_t = (Bz)_t,$$

and thus

$$\begin{aligned} \sum_{j=1}^{n} (b_{tj} - a_{tj})\frac{z_j}{z_t} + \alpha &= \frac{(Bz)_t}{z_t} \\ &= f_B(z) \\ &\leq \beta. \end{aligned} \tag{9}$$

Now, $z \in E^n$, and therefore

$$\sum_{j=1}^{n} (b_{tj} - a_{tj}) z_j \geq \min_j (b_{tj} - a_{tj})$$
$$\geq \min_{i,j} (b_{ij} - a_{ij})$$
$$= m. \tag{10}$$

From (9) and (10) it follows that

$$\beta - \alpha \geq \frac{m}{z_t}.$$

It remains to show that

$$z_t \leq \max_{i,j} (s_{ij}/c_j). \tag{11}$$

Observe that z is an eigenvector of S. For,

$$ASz = SAz$$
$$= \alpha Sz,$$

and thus Sz is a maximal eigenvector of A, and therefore it must be a positive multiple of z,

$$Sz = \lambda z, \tag{12}$$

for some $\lambda > 0$. But then, by formula (3), Section 2.1,

$$\lambda = \sum_{j=1}^{n} c_j z_j.$$

On the other hand, from (12)

$$\lambda z_t = \sum_{j=1}^{n} s_{tj} z_j,$$

and therefore

$$z_t = \frac{1}{\lambda} \sum_{j=1}^{n} s_{tj} z_j$$
$$= \frac{\sum_{j=1}^{n} s_{tj} z_j}{\sum_{j=1}^{n} c_j z_j}$$
$$\leq \max_j \frac{s_{tj}}{c_j},$$

by (7), Section 2.1. Inequality (11) follows immediately. The upper inequality is proved similarly [this time using the function g_A (see Theorems 4.5 and 4.6, Chapter I)]. ■

2.3. BOUNDS FOR MAXIMAL EIGENVECTORS

Let A be a positive $n \times n$ matrix, r its maximal eigenvalue, and $x = (x_1, x_2, \ldots, x_n)$ a maximal eigenvector of A. The problem of localizing r is related to the problem of estimating the quotient $\max_{i,j}(x_i/x_j)$. The lower bound in the following theorem is essentially due to Ledermann [3]. The upper bound is Minc's [6] improvement of a result of Ostrowski [8].

Theorem 3.1. *Let $A = (a_{ij})$ be a positive matrix with a maximal eigenvector $x = (x_1, x_2, \ldots, x_n)$ and let $\gamma = \max_{i,j}(x_i/x_j)$. Then*

$$\sqrt{\frac{R}{\rho}} \le \gamma \le \max_{j,s,t} \frac{a_{sj}}{a_{tj}}, \tag{1}$$

where R and ρ are the greatest and least row sums of A, respectively. The left inequality in (1) is an equality if and only if $R = \rho$. Equality holds on the right-hand side of (1) if and only if the pth row of A is a multiple of the qth row, for some pair of indices p and q satisfying $a_{ph}/a_{qh} = \max_{j,s,t}(a_{sj}/a_{tj})$.

Proof. Let $x_m = \max_i x_i$ and $x_\mu = \min_i x_i$, and let r be the maximal eigenvalue of A. Then

$$\begin{aligned} rx_i &= \sum_{j=1}^{n} a_{ij}x_j \\ &\ge \sum_{j=1}^{n} a_{ij}x_\mu \\ &= r_i x_\mu, \qquad i = 1, 2, \ldots, n. \end{aligned} \tag{2}$$

Therefore

$$\frac{r}{r_i} \ge \frac{x_\mu}{x_i} \ge \frac{1}{\gamma},$$

for all i. In particular,

$$\frac{r}{R} \ge \frac{1}{\gamma}. \tag{3}$$

Similarly,

$$rx_i \le r_i x_m,$$

and therefore

$$\frac{r}{r_i} \le \frac{x_m}{x_i} \le \gamma,$$

for all i, and, in particular,

$$\frac{r}{\rho} \leq \gamma. \tag{4}$$

From (3) and (4), it follows that

$$\gamma \geq \sqrt{\frac{R}{\rho}}. \tag{5}$$

Let $r_m = R$. Equality can hold in (5) only if it holds in (2), that is,

$$rx_m = \sum_{j=1}^{n} a_{mj}x_j = Rx_\mu.$$

This implies that all the coordinates of x are equal. But then all the row sums of A must be equal. The converse is obvious.

To prove the upper bound we note again that

$$rx_m = \sum_{j=1}^{n} a_{mj}x_j, \tag{6}$$

and

$$rx_\mu = \sum_{j=1}^{n} a_{\mu j}x_j. \tag{7}$$

Dividing (6) by (7), we have

$$\gamma = \frac{x_m}{x_\mu} = \frac{\sum_{j=1}^{n} a_{mj}x_j}{\sum_{j=1}^{n} a_{\mu j}x_j}.$$

Applying inequality (7), Section 2.1, we obtain

$$\gamma \leq \max_j \frac{a_{mj}x_j}{a_{\mu j}x_j} \tag{8}$$

$$= \max_j \frac{a_{mj}}{a_{\mu j}}$$

$$\leq \max_{j,s,t} \frac{a_{sj}}{a_{tj}}. \tag{9}$$

Suppose now that the right-hand side of (1) is an equality. Then equality must hold both in (8) and (9). Now, (8) is an equality (see Section 2.1) if and only if

$$\frac{a_{mj}}{a_{\mu j}} = \gamma,$$

for $j = 1, 2, \ldots, n$. Hence the mth and the μth rows of A are linearly dependent. Moreover, since equality holds in (9), for any h we have

$$\frac{a_{mh}}{a_{\mu h}} = \max_{j,s,t} \frac{a_{sj}}{a_{tj}}.$$

Conversely, suppose that $a_{pj}/a_{qj} = \beta$, $j = 1, 2, \ldots, n$, and that $a_{pj}/a_{qj} \geq a_{sj}/a_{tj}$ for all s, t, and j. Then

$$\frac{x_p}{x_q} = \frac{\sum_{j=1}^n a_{pj} x_j}{\sum_{j=1}^n a_{qj} x_j} = \beta,$$

and for any s and t,

$$\frac{x_s}{x_t} = \frac{\sum_{j=1}^n a_{sj} x_j}{\sum_{j=1}^n a_{tj} x_j} \leq \max_j \frac{a_{sj}}{a_{tj}} \leq \beta = \frac{x_p}{x_q}.$$

Hence

$$\frac{x_p}{x_q} = \max_{i,j} \frac{x_i}{x_j},$$

and therefore

$$\gamma = \beta = \max_{j,s,t} \frac{a_{sj}}{a_{tj}},$$

that is, the right-hand side of (1) is an equality. This concludes the proof of the theorem. ■

Note that since A and $A + \varepsilon I_n$ have common eigenvectors, the upper bound in (1) may be improved by adding to A a suitable multiple of I_n. Also, by use of the same device, the theorem can be extended to the case of nonnegative matrices all of whose zero entries (if any) are on the main diagonal (see [6], page 108).

Example 3.1. Find bounds for γ for the matrix

$$A = \begin{bmatrix} 2 & 1 & 2 \\ 3 & 2 & 3 \\ 2 & 1 & 2 \end{bmatrix}.$$

Using Theorem 3.1 we obtain the inequalities

$$1.2649\ldots = \sqrt{\tfrac{8}{5}} < \gamma < 2. \tag{10}$$

Both bounds in (10) can be sharpened by subtracting from A an appropriate multiple of I_3 and applying Theorem 3.1 to the new matrix. As noted above, a maximal eigenvector of A is also a maximal eigenvector of $A - \varepsilon I_3$. If we let ε tend to 2 from the left, then the lower bound approaches $\sqrt{6/3} = 1.4142\ldots$. In order to improve the upper bound in (10), clearly we must have $0 < \varepsilon < \frac{1}{2}$. Then

$$\gamma < \max\left(\frac{3}{2-\varepsilon}, 2-\varepsilon\right).$$

Hence the least upper bound is obtained, if

$$\frac{3}{2-\varepsilon} = 2 - \varepsilon,$$

that is, if

$$\varepsilon = 2 - \sqrt{3}.$$

Then Theorem 3.1 yields the upper bound

$$\gamma < \sqrt{3} = 1.73205\ldots.$$

The actual value of γ, computed directly, is $\sqrt{7} - 1 = 1.64575\ldots$. ■

Example 3.2. Find bounds for γ for the matrix

$$B = \begin{bmatrix} 2 & 1 & 1 \\ 1 & 0 & 1 \\ 1 & 1 & 0 \end{bmatrix}.$$

Note that B is irreducible, and therefore γ is well defined. Apply Theorem 3.1 to the matrix $B + \varepsilon I_3$, where $0 < \varepsilon < 1$:

$$\sqrt{\frac{4+\varepsilon}{2+\varepsilon}} < \gamma < \max\left\{\frac{2+\varepsilon}{1}, \frac{1}{\varepsilon}\right\}. \qquad (11)$$

Clearly, (1) yields the best upper bound for γ when $2 + \varepsilon = 1/\varepsilon$, that is, when $\varepsilon = -1 + \sqrt{2}$. Hence $\gamma \le 1 + \sqrt{2}$.

The lower bound in (11) is less than $\sqrt{2}$, but it tends to $\sqrt{2}$ as ε tends to 0. We can conclude therefore that $\gamma \ge \sqrt{2}$. The actual value of γ for the matrix B is 2. ■

PROBLEMS

1 Show that the condition for equality in Theorem 1.1 is not necessary, if the matrix A is reducible.

2 Let A be an irreducible matrix, and let r, R, and ρ be as defined in the statement of Theorem 1.1. Show that $\rho \le r \le R$ by relating r to $f_A(u)$ and $g_A(u)$ for an appropriate vector u.

3 The matrix

$$A = \begin{bmatrix} 1 & 2 & 2 & 2 \\ 2 & 1 & 1 & 2 \\ 1 & 1 & 1 & 1 \\ 2 & 2 & 2 & 1 \end{bmatrix}$$

has a maximal eigenvector $(12, 11, 7, 12)$. Find a positive matrix similar to A, with maximal eigenvector $(12, 12, 11, 7)$.

4 Let $A = (a_{ij})$ be a positive $n \times n$ matrix with maximal eigenvalue r, and row sums $r_1, r_2, \ldots, r_n$. Let $x = (x_1, x_2, \ldots, x_n)$, where $x_1 \geq x_2 \geq \cdots \geq x_n > 0$, be a maximal eigenvector of A. If $x_1 > x_n$ show that $r_n < r < r_1$.

5 Let

$$A = \begin{bmatrix} 3 & 3 & 1 \\ 1 & 3 & 1 \\ 1 & 3 & 3 \end{bmatrix}.$$

(a) Compute the bounds for the maximal eigenvalue r of A using the formulas in Theorems 1.1–1.5. Find r by direct computation and compare it with the bounds obtained.

(b) Use the method in Example 1.2 to obtain improved bounds for r.

(c) Obtain bounds for r using the method in Example 1.3.

6 Find bounds for γ for the matrix A in Problem 5, using Theorem 3.1. Compute a maximal eigenvector of A, evaluate γ directly, and compare the value found with the bounds.

7 Let A be the matrix in Problem 5. Find the value ε for which Theorem 3.1 applied to $\varepsilon I_3 + A$ yields the best upper bound for γ.

8 Let

$$B = \begin{bmatrix} 1 & 1 & 1 & 1 \\ 1 & 1 & 2 & 2 \\ 1 & 2 & 1 & 2 \\ 1 & 2 & 2 & 3 \end{bmatrix}.$$

Compute the bounds for the maximal eigenvalue of B using Theorems 1.1–1.5.

9 Compute the bounds for γ for the matrix B in Problem 8, using Theorem 3.1. Improve the upper bound by applying the theorem to matrix $B + \varepsilon I_4$ for an appropriate ε.

10 Let A be a nonnegative matrix with maximal eigenvalue r. Suppose that A dominates a complex matrix C with eigenvalue s and that $|s| = r$. Show by a counterexample that the condition (2) in Theorem 2.1 does not necessarily hold.

11 Illustrate the result in Theorem 2.1 for the matrix A in Problem 5 and

$$C = \begin{bmatrix} 3i & 3 & (-1+i)/\sqrt{2} \\ -1 & 3i & (-1-i)/\sqrt{2} \\ (1+i)/\sqrt{2} & (3-3i)/\sqrt{2} & 3i \end{bmatrix}.$$

12 Let A be the matrix in Problem 5 and let $B = I_3 + A$. Use Theorem 2.2 to find the upper bound for the difference of the maximal eigenvalues of B and A, taking (i) $S = A$, (ii) $S = A^2$.

13 In Problem 12 take $S = \theta I_3 + A$ and find the value θ for which Theorem 2.2 yields the best upper bound.

14 Let $A = (a_{ij})$ be a positive $n \times n$ matrix. Show that

$$\gamma \geq \sqrt{\frac{R-\delta}{\rho-\delta}},$$

where γ, R, and ρ are as defined in Theorem 3.1, and $\delta = \min_i a_{ii}$.

15 Prove in detail the upper bound in Theorem 2.2.

REFERENCES

1. A. Brauer, The theorems of Ledermann and Ostrowski on positive matrices, *Duke Math. J.* **24** (1957), 265–274.
2. G. Frobenius, Über Matrizen aus nicht negativen Elementen, *S.-B. K. Preuss. Akad. Wiss. Berlin* (1912), 456–477.
3. W. Ledermann, Bounds for the greatest latent root of a positive matrix, *J. London Math. Soc.* **25** (1950), 265–268.
4. M. Marcus and H. Minc, *Modern University Algebra*, Macmillan, New York, 1965.
5. M. Marcus, H. Minc, and B. Moyls, Some results on nonnegative matrices, *J. Res. Nat. Bur. Standards Sect. B* **65** (1961), 205–209.
6. H. Minc, On the maximal eigenvector of a positive matrix, *SIAM J. Numer. Anal.* **7** (1970), 424–427.
7. A. Ostrowski, Bounds for the greatest latent root of a positive matrix, *J. London Math. Soc.* **27** (1952), 253–256.
8. A. M. Ostrowski, On the eigenvector belonging to the maximal root of a non-negative matrix, *Proc. Edinburgh Math. Soc.* **12** (1960/61), 107–112.
9. H. Wielandt, Unzerlegbare nicht negative Matrizen, *Math. Z.* **52** (1950), 642–648.

III

Primitive and Imprimitive Matrices

3.1. SPECTRA OF IRREDUCIBLE MATRICES

In Chapter I we extended Perron's results for positive matrices to the larger class of nonnegative matrices. The first three sections of the present chapter contain further results of Frobenius [1] on spectral properties of nonnegative matrices which have no counterpart in the theory of positive matrices.

Definition 1.1. Let A be an irreducible $n \times n$ matrix with maximal eigenvalue r, and suppose that A has exactly h eigenvalues of modulus r. The number h is called the *index of imprimitivity* of A, or simply the *index* of A. If $h = 1$, then the matrix A is said to be *primitive*; otherwise, it is *imprimitive* (or *cyclic*, in some authors' nomenclature).

The proofs of the following two theorems are due to Wielandt [9].

Theorem 1.1. *Let A be an irreducible $n \times n$ matrix with maximal eigenvalue r and index h. Let $\lambda_1, \lambda_2, \ldots, \lambda_h$ be the eigenvalues of A of modulus r. Then $\lambda_1, \lambda_2, \ldots, \lambda_h$ are the distinct hth roots of r^h.*

Proof. Let $\lambda_t = re^{i\varphi_t}$, $t = 1, 2, \ldots, h$. Since $|\lambda_t| = r$, the equality condition in Theorem 2.1, Chapter II, with $C = A$ and $s = \lambda_t$, implies that

$$A = e^{i\varphi_t} D_t A D_t^{-1}, \qquad t = 1, 2, \ldots, h. \tag{1}$$

Hence A and $e^{i\varphi_t}A$ are similar. Since r is a simple eigenvalue of A, it follows that, for each t, $e^{i\varphi_t}r = \lambda_t$ is a simple eigenvalue of $e^{i\varphi_t}A$, and thus of A. Now, by (1),

$$\begin{aligned} A &= e^{i\varphi_t} D_t \left(e^{i\varphi_s} D_s A D_s^{-1} \right) D_t^{-1} \\ &= e^{i(\varphi_t + \varphi_s)} (D_t D_s) A (D_t D_s)^{-1}, \end{aligned}$$

and therefore A and $e^{i(\varphi_t+\varphi_s)}A$ are similar for any s and t. We can conclude that $re^{i(\varphi_t+\varphi_s)}$ is an eigenvalue of A, and therefore $e^{i(\varphi_t+\varphi_s)}$ must be one of the numbers $e^{i\varphi_1}, e^{i\varphi_2}, \ldots, e^{i\varphi_h}$. Hence the h distinct numbers $e^{i\varphi_1}, e^{i\varphi_2}, \ldots, e^{i\varphi_h}$ are closed under multiplication, and therefore they are the hth roots of unity. ■

Theorem 1.2. *The spectrum of an irreducible matrix of index h is invariant under a rotation through $2\pi/h$, but not through a positive angle smaller than $2\pi/h$.*

Proof. Let the spectrum of A be $\sigma = (\lambda_1, \lambda_2, \ldots, \lambda_n)$. Then the spectrum of $e^{i2\pi/h}A$ is $\tau = (\lambda_1 e^{i2\pi/h}, \lambda_2 e^{i2\pi/h}, \ldots, \lambda_n e^{i2\pi/h})$, that is, the spectrum σ rotated through $2\pi/h$. It was shown in the proof of the preceding theorem that the matrices A and $e^{i2\pi/h}A$ are similar, and therefore τ is the spectrum of A as well. This proves the first part of the theorem. It is clear from the result in Theorem 1.1 that a rotation through any angle smaller than $2\pi/h$ cannot hold the spectrum fixed, since even the set of the eigenvalues of maximum modulus is not then preserved. ■

The following theorem relates the index of an imprimitive matrix to the form of its characteristic polynomial. It is often used in determining whether an irreducible matrix is primitive or not.

Theorem 1.3. *Let A be an irreducible matrix with index h, and let*

$$\lambda^n + a_1\lambda^{n_1} + a_2\lambda^{n_2} + \cdots + a_k\lambda^{n_k},$$

where $n > n_1 > n_2 > \cdots > n_k$ and $a_t \neq 0$, $t = 1, 2, \ldots, k$, be the characteristic polynomial of A. Then

$$h = \text{g.c.d.}(n - n_1, n_1 - n_2, \ldots, n_{k-1} - n_k). \tag{2}$$

Proof. Suppose that $m \geq 2$ is an integer such that A and $e^{i2\pi/m}A$ have the same spectrum. Then their characteristic polynomials are equal:

$$\begin{aligned} &\lambda^n + a_1\lambda^{n_1} + a_2\lambda^{n_2} + \cdots + a_k\lambda^{n_k} \\ &\quad = \lambda^n + a_1\theta^{n-n_1}\lambda^{n_1} + a_2\theta^{n-n_2}\lambda^{n_2} + \cdots + a_k\theta^{n-n_k}\lambda^{n_k}, \end{aligned}$$

where $\theta = e^{i2\pi/m}$. It follows that

$$a_t = a_t\theta^{n-n_t},$$

for $t = 1, 2, \ldots, k$, and therefore m divides each of $n - n_1, n - n_2, \ldots, n - n_k$. Now, by Theorem 1.2, the matrices A and $e^{i2\pi/m}A$ have the same spectrum for $m = h$ but not for $m > h$. This, together with the preceding argument, implies that

$$h = \text{g.c.d.}(n - n_1, n - n_2, \ldots, n - n_k),$$

and (2) follows immediately. ■

Corollary 1.1. *An irreducible matrix with positive trace is primitive.*

Indeed, by Theorem 1.3, the index of the matrix is

$$\text{g.c.d.}(n - (n - 1), \ldots) = 1.$$

3.2. PRIMITIVE MATRICES

It was shown in Corollary 1.1 that an irreducible matrix with a positive trace is primitive. In particular, if A is irreducible, then $I_n + A$ is primitive. In Section 1.2 we discovered that the matrix $I_n + A$ has the remarkable property that it becomes positive when raised to sufficiently high power (see Corollary 2.2, Chapter I). We now show that this is a general property of all primitive matrices. In fact, it is an alternative definition of primitive matrices.

Theorem 2.1 (Frobenius [1]). *A necessary and sufficient condition for a nonnegative matrix A to be primitive is that A^m be positive for some positive integer m.*

Proof (Marcus and Minc [2]). Suppose that $A^m > 0$. Then A must be irreducible. For, if A were reducible, that is, if A were cogredient to a matrix of the form

$$\begin{bmatrix} B & C \\ 0 & D \end{bmatrix},$$

then A^m would be cogredient to a matrix of the form

$$\begin{bmatrix} B^m & C' \\ 0 & D^m \end{bmatrix},$$

and thus could not be positive. We now show that h, the index of A, is 1. Let r be the maximal eigenvalue of A and let $re^{i2\pi t/h}$, $t = 0, 1, \ldots, h - 1$, be the eigenvalues of A of modulus r (see Theorem 1.1). Then A^m has h eigenvalues of modulus r^m. Since A^m is positive it is primitive, by Corollary 1.1, and thus h must be 1.

Conversely, let A be a primitive matrix with maximal root r. Then the matrix A/r is primitive as well, its maximal root is 1, and all its other roots have moduli less than 1. Let

$$S^{-1}\left(\frac{1}{r}A\right)S = 1 \dotplus B \tag{1}$$

be the Jordan normal form of A/r. We can deduce immediately from (1) that

(i) the moduli of all roots of B are less than 1, and therefore $\lim_{t\to\infty} B^t = 0$ (Problem 1);

(ii) the first column of S is a characteristic vector of A/r corresponding to the maximal root 1, and therefore has no zero coordinates;

(iii) the first row of S^{-1} is a characteristic vector of the transpose of A/r corresponding to its maximal root, and thus cannot have zero coordinates.

Now,

$$\begin{aligned}\lim_{t\to\infty}\left(\frac{1}{r}A\right)^t &= \lim_{t\to\infty}\left(S(1 \dotplus B)S^{-1}\right)^t \\ &= S\left(1 \dotplus \left(\lim_{t\to\infty} B^t\right)\right)S^{-1} \\ &= S(1 \dotplus 0)S^{-1}\end{aligned}$$

is a nonnegative matrix. But the (i, j) entry of $S(1 \dotplus 0)S^{-1}$ is the nonzero product $S_{i1}(S^{-1})_{1j}$ for all i and j. Hence $S(1 \dotplus 0)S^{-1}$ must be strictly positive, that is,

$$\lim_{t\to\infty}\left(\frac{1}{r}A\right)^t > 0.$$

It follows that for a sufficiently large integer m,

$$\left(\frac{1}{r}A\right)^m > 0,$$

and therefore

$$A^m > 0. \quad \blacksquare$$

It should be emphasized that the result in Theorem 2.1 is not obvious per se, since a product of primitive matrices need not even be irreducible. On the other hand, a product of reducible matrices may be positive. For example, both

$$\begin{bmatrix} 1 & 1 \\ 1 & 0 \end{bmatrix} \quad \text{and} \quad \begin{bmatrix} 0 & 1 \\ 1 & 1 \end{bmatrix}$$

are primitive, but

$$\begin{bmatrix} 1 & 1 \\ 1 & 0 \end{bmatrix}\begin{bmatrix} 0 & 1 \\ 1 & 1 \end{bmatrix} = \begin{bmatrix} 1 & 2 \\ 0 & 1 \end{bmatrix}$$

is reducible. The second assertion is quite obvious.

3.3. THE FROBENIUS FORM OF AN IRREDUCIBLE MATRIX

Frobenius [1] discovered a remarkable connection between the spectral properties of an irreducible matrix and its zero pattern, that is, the distribution of its zero entries.

Theorem 3.1. *Let A be an irreducible matrix with index $h \geq 2$. Then A is cogredient to a matrix of the form*

$$\begin{bmatrix} 0 & A_{12} & 0 & \cdots & 0 & 0 \\ 0 & 0 & A_{23} & \cdots & 0 & 0 \\ \vdots & & & \ddots & & \vdots \\ 0 & 0 & & \cdots & 0 & A_{h-1,h} \\ A_{h1} & 0 & & \cdots & & 0 \end{bmatrix}, \tag{1}$$

where the zero blocks along the main diagonal are square.

Proof (Wielandt [9]). Let r be the maximal eigenvalue of A. Then, as we saw in Theorem 1.1,

$$\lambda_t = re^{i2\pi t/h}, \qquad t = 0, 1, \ldots, h-1,$$

are the eigenvalues of A of modulus r, and

$$A = e^{i2\pi t/h} D_t A D_t^{-1}, \qquad t = 0, 1, \ldots, h-1, \tag{2}$$

where $|D_t| = I_n$. We can assume without loss of generality that the (1, 1) entry of each D_t is 1. Let z be a positive eigenvector of A corresponding to r. Define

$$z^t = D_t z, \qquad t = 0, 1, \ldots, h-1. \tag{3}$$

By (2) and (3),

$$\begin{aligned} Az^t &= e^{i2\pi t/h} D_t A D_t^{-1} D_t z \\ &= e^{i2\pi t/h} D_t rz \\ &= \lambda_t z^t, \end{aligned}$$

and therefore z^t is an eigenvector of A corresponding to λ_t. Since the eigenspace of each λ_t is one-dimensional, the z^t (and thus the D_t), $t = 0, 1, \ldots, h-1$, are determined within a constant. But the first coordinate of each D_t is 1, and therefore the D_t are uniquely determined. Now, applying (2) twice, we have

$$\begin{aligned} A &= e^{i2\pi t/h} D_t \left(e^{i2\pi s/h} D_s A D_s^{-1} \right) D_t^{-1} \\ &= e^{i2\pi(s+t)/h} (D_t D_s) A (D_t D_s)^{-1}, \end{aligned}$$

and we can show, by reasoning analogous to that following (3), that $D_t D_s z$ is an eigenvector corresponding to $re^{i2\pi(s+t)/h}$. In particular, $D_1^h z$ is an eigenvector corresponding to $re^{i2\pi h/h} = r$. By the uniqueness of the D_t, we can conclude that

$$D_1^h = I_n,$$

so that the main diagonal entries of D_1 are hth roots of 1. Let P be a permutation matrix such that

$$P^{\mathrm{T}} D_1 P = \sum_{t=1}^{s} {}^{\cdot}\, e^{im_t 2\pi/h} I_{n_t},$$

where $0 = m_1 < m_2 < \cdots < m_s \leq h - 1$. Partition $P^{\mathrm{T}}AP$ into blocks conformally to the partition of $P^{\mathrm{T}}D_1P$ above; that is, let

$$P^{\mathrm{T}}AP = \begin{bmatrix} A_{11} & A_{12} & \cdots & A_{1s} \\ A_{21} & A_{22} & \cdots & A_{2s} \\ \vdots & & \ddots & \vdots \\ A_{s1} & A_{s2} & \cdots & A_{ss} \end{bmatrix}, \tag{4}$$

where block A_{pq} is $n_p \times n_q$, $p, q = 1, 2, \ldots, s$. Equating the (p, q) blocks on both sides of $P^{\mathrm{T}}AP = e^{i2\pi/h}(P^{\mathrm{T}}D_1P)(P^{\mathrm{T}}AP)(P^{\mathrm{T}}D_1^{-1}P)$, we obtain

$$A_{pq} = e^{i(1+m_p-m_q)2\pi/h} A_{pq}.$$

Therefore for each (p, q), either

$$A_{pq} = 0, \tag{5}$$

or

$$m_q - m_p \equiv 1 \pmod{h}. \tag{6}$$

The matrix A is irreducible, and thus for no p can A_{pq} be 0 for all q, nor for any q can A_{pq} vanish for all p.

If $p = 1$, then the congruence (6) is

$$m_q \equiv 1 \pmod{h}, \tag{7}$$

and since $1 \leq m_2 < m_3 < \cdots < m_s \leq h - 1$, the only solution of (7) is $m_2 = 1$. Thus $A_{1q} = 0$ for all $q \neq 2$. Next, for $p = 2$ condition (6) becomes

$$m_q - m_2 \equiv 1 \pmod{h},$$

that is,

$$m_q \equiv 2 \pmod{h},$$

and as above we find that $m_3 = 2$, and $A_{2q} = 0$ for all $q \neq 3$. We continue in the same manner and conclude that $m_{p+1} = p$, and $A_{pq} = 0$ for all $q \neq p+1$, $p = 1, 2, \ldots, s-1$.

Consider the case $p = s$. Conditions (5) and (6) state that for each q either $A_{sq} = 0$ or $m_q - m_s \equiv 1 \pmod{h}$, that is, $m_q \equiv s \pmod{h}$. Now, A_{s1} cannot be 0, since all other A_{p1} are 0. Hence we must have

$$m_1 \equiv s \pmod{h},$$

which implies that $s = h$, and thus $m_q \not\equiv s \pmod{h}$ for all $q \neq 1$. It follows that $A_{sq} = 0$ for all $q \neq 1$. This concludes the proof. ■

Many spectral properties of imprimitive matrices can be deduced immediately from the Frobenius form of the matrices. Apart from the index of imprimitivity h of the matrix, it is obvious from the form (1) that the trace of an imprimitive matrix must be zero, and if $h > 2$, then the second symmetric function of its eigenvalues must also vanish (see Theorem 1.3). Moreover, certain structural properties of imprimitive matrices are quite obvious: Some power of the matrix must be cogredient (and thus similar) to a direct sum. In the next section we shall study matrices in the Frobenius form, or in any superdiagonal block form, and their spectra.

Example 3.1. Use the result in Theorem 3.1 to prove that the spectrum of an irreducible matrix A with index h is invariant under a rotation through $2\pi/h$.

Let P be a permutation matrix such that $P^{\mathrm{T}}AP$ is in the form (1). Let

$$D = \sum_{t=1}^{h}{}^{\cdot}\, e^{i2\pi t/h} I_{n_t}.$$

Then

$$D^{-1}(P^{\mathrm{T}}AP)D = e^{i2\pi/h}P^{\mathrm{T}}AP.$$

Thus A and $e^{i2\pi/h}A$ are similar. The proof is complete. ■

3.4. MATRICES IN SUPERDIAGONAL BLOCK FORM

A matrix A in the form

$$\begin{bmatrix} 0 & A_{12} & 0 & \cdots & 0 & 0 \\ 0 & 0 & A_{23} & \cdots & 0 & 0 \\ \vdots & & & \ddots & & \vdots \\ 0 & 0 & & \cdots & 0 & A_{k-1,k} \\ A_{k1} & 0 & & \cdots & & 0 \end{bmatrix}, \tag{1}$$

where the block $A_{i,i+1}$ is $n_i \times n_{i+1}$, $i = 1, 2, \ldots, k-1$, and A_{k1} is $n_k \times n_1$, is said to be in a *superdiagonal block form*, or more specifically, in the *superdiagonal* $(n_1, n_2, \ldots, n_k)$*-block form*. If A is irreducible with index k, then clearly (1) is the *Frobenius form* of A.

A matrix in the form (1) is not necessarily irreducible, even if all the blocks $A_{i,i+1}$ happen to be square and primitive. For example, the matrix

$$\left[\begin{array}{cc|cc} \multicolumn{2}{c|}{0} & 0 & 1 \\ & & 1 & 1 \\ \hline 1 & 1 & \multicolumn{2}{c}{0} \\ 1 & 0 & & \end{array}\right]$$

is reducible although its (1, 2) and (2, 1) blocks are primitive. On the other hand, a matrix may have the form (1), where all the blocks are reducible, and still be irreducible. For example,

$$\left[\begin{array}{cc|cc} \multicolumn{2}{c|}{0} & 1 & 0 \\ & & 1 & 1 \\ \hline 1 & 1 & \multicolumn{2}{c}{0} \\ 0 & 1 & & \end{array}\right]$$

is irreducible although its (1, 2) and (2, 1) blocks are reducible. In this section we obtain necessary and sufficient conditions for a nonnegative matrix in the form (1) to be irreducible.

We shall require the following lemmas.

Lemma 4.1. *If A is in a superdiagonal $(n_1, n_2, \ldots, n_k)$-block form,*

$$A^k = \sum_{j=1}^{k} B_j,$$

where $B_j = A_{j,j+1}A_{j+1,j+2} \cdots A_{j-1,j}$ (subscripts reduced modulo k) is n_j-square, $j = 1, 2, \ldots, k$.

Lemma 4.2. *All the matrices B_j defined in Lemma* 4.1 *have the same nonzero eigenvalues.*

Lemma 4.1 is immediate. The result in Lemma 4.2 is essentially due to Sylvester [8]; the proof is quite straightforward (Problem 5).

If σ and τ are permutations on $\{1, 2, \ldots, n\}$, we define the composite permutation $\sigma\tau$ by

$$\sigma\tau(i) = \sigma(\tau(i)), \qquad i = 1, 2, \ldots, n.$$

The $n \times n$ permutation matrix $A(\sigma)$ is called the *incidence matrix* of σ if its (i, j) entry is $\delta_{i, \sigma(j)}$. We have then

$$A(\sigma\tau) = A(\sigma)A(\tau).$$

Theorem 4.1 (Minc [4]). *Let A be an irreducible matrix with index h. Then A is cogredient to a matrix in the form* (1) *with k nonzero blocks if and only if k divides h.*

Proof. Suppose that A is cogredient to a matrix in the form (1). Then A^k is cogredient to

$$\sum_{j=1}^{k} {}^{\cdot} B_j,$$

where the B_j are as defined in Lemma 4.1. Let r be the maximal eigenvalue of A. Then A^k has exactly h eigenvalues of modulus r^k, and all its other eigenvalues have moduli smaller than r^k. Now, by Lemma 4.2, the h eigenvalues of maximum modulus must be equally divided among the blocks $B_1, B_2, \ldots, B_k$, and thus each of the B_j must have exactly h/k eigenvalues of modulus r^k. It follows that k must divide h.

Conversely, let A be an irreducible matrix with index of imprimitivity h, and let k be any divisor of h. Let Q be a permutation matrix such that QAQ^{T} is in the superdiagonal $(n_1, n_2, \ldots, n_h)$-block form, as in (1), Section 3.3. We show that there exists a permutation matrix P such that $(PQ)A(PQ)^{\mathrm{T}}$ is in the form (1) with k nonzero blocks in the superdiagonal.

Let τ be the cycle $(h, h-1, \ldots, 2, 1)$ so that $A(\tau)$ is the $h \times h$ permutation matrix with 1's in positions $(i, i+1)$, $i = 1, 2, \ldots, h-1$, and $(h, 1)$. Let $h = km$, and let $\sigma \in S_h$ be defined by

$$\sigma(sk + t) = s + 1 + (t-1)m, \qquad s = 0, 1, \ldots, m-1,\ t = 1, 2, \ldots, k,$$

that is,

$$\sigma = \begin{pmatrix} 1 & 2 & \cdots & k & k+1 & k+2 & \cdots & 2k & 2k+1 & \cdots & km \\ 1 & 1+m & \cdots & 1+(k-1)m & 2 & 2+m & \cdots & 2+(k-1)m & 3 & \cdots & km \end{pmatrix}.$$

It follows that

$$\sigma\tau\sigma^{-1}(i) = i - m, \qquad i = m+1, m+2, \ldots, h,$$
$$\sigma\tau\sigma^{-1}(1) = h,$$

and

$$\sigma\tau\sigma^{-1}(i) = h - m + i - 1,$$

for $i = 2, 3, \ldots, m$. Thus

$$A(\sigma)A(\tau)A(\sigma)^{\mathrm{T}} = \begin{bmatrix} 0 & I_{h-m} \\ P_m & 0 \end{bmatrix}, \tag{2}$$

where P_m denotes the $m \times m$ permutation matrix with 1's in the positions $(1,2), (2,3), \ldots, (m-1, m)$, and $(m, 1)$.

We now augment matrices $A(\sigma)$ and $A(\tau)$ to $n \times n$ matrices P and QAQ^{T}, respectively, by replacing the 1's in $A(\sigma)$ by identity matrices, the 1's in $A(\tau)$ by blocks $A_{i,i+1}$, and all the zero entries by zero blocks of appropriate size. Then $PQAQ^{\mathrm{T}}P^{\mathrm{T}}$ is an $n \times n$ matrix in a superdiagonal block form obtained from matrix (2) by replacing the h 1's by appropriate blocks $A_{i,i+1}$. Specifically, let P be the $n \times n$ permutation matrix partitioned into blocks so that the $(i, \sigma^{-1}(i))$ block is $I_{n_{\sigma^{-1}(i)}}$, $i = 1, 2, \ldots, h$. Then the matrix $PQAQ^{\mathrm{T}}P^{\mathrm{T}}$ is in the superdiagonal $(g_1, g_2, \ldots, g_k)$-block form, where

$$g_t = \sum_{j=m(t-1)+1}^{mt} n_{\sigma^{-1}(j)}, \qquad t = 1, 2, \ldots, k. \quad \blacksquare$$

Example 4.1. We illustrate Theorem 4.1 by considering a matrix A such that QAQ^{T} is in the Frobenius form with $h = 12$, where the main diagonal blocks are of orders $n_1, n_2, \ldots, n_{12}$. Find a permutation matrix P so that $PQAQ^{\mathrm{T}}P^{\mathrm{T}}$ is in a superdiagonal block form with $k = 4$.

Let

$$\tau = (12, 11, 10, 9, 8, 7, 6, 5, 4, 3, 2, 1),$$

and

$$\sigma = \begin{pmatrix} 1 & 2 & 3 & 4 & 5 & 6 & 7 & 8 & 9 & 10 & 11 & 12 \\ 1 & 4 & 7 & 10 & 2 & 5 & 8 & 11 & 3 & 6 & 9 & 12 \end{pmatrix}.$$

Then

$$\sigma\tau\sigma^{-1} = \begin{pmatrix} 1 & 2 & 3 & 4 & 5 & 6 & 7 & 8 & 9 & 10 & 11 & 12 \\ 12 & 10 & 11 & 1 & 2 & 3 & 4 & 5 & 6 & 7 & 8 & 9 \end{pmatrix},$$

$$A(\sigma)A(\tau)A(\sigma)^{\mathrm{T}} = \left[\begin{array}{ccc|c} & 0 & & I_9 \\ \hline 0 & 1 & 0 & \\ 0 & 0 & 1 & 0 \\ 1 & 0 & 0 & \end{array}\right],$$

and thus

$PQAQ^{\mathrm{T}}P^{\mathrm{T}} =$

$$\left[\begin{array}{c|c|c|c} 0 & \begin{matrix} A_{12} & 0 & 0 \\ 0 & A_{56} & 0 \\ 0 & 0 & A_{9,10} \end{matrix} & 0 & 0 \\ \hline 0 & 0 & \begin{matrix} A_{23} & 0 & 0 \\ 0 & A_{67} & 0 \\ 0 & 0 & A_{10,11} \end{matrix} & 0 \\ \hline 0 & 0 & 0 & \begin{matrix} A_{34} & 0 & 0 \\ 0 & A_{78} & 0 \\ 0 & 0 & A_{11,12} \end{matrix} \\ \hline \begin{matrix} 0 & A_{45} & 0 \\ 0 & 0 & A_{89} \\ A_{12,1} & 0 & 0 \end{matrix} & 0 & 0 & 0 \end{array}\right]$$

where

$$P = \begin{bmatrix} I_{n_1} & 0 & 0 & 0 & 0 & 0 & 0 & 0 & 0 & 0 & 0 & 0 \\ 0 & 0 & 0 & 0 & I_{n_5} & 0 & 0 & 0 & 0 & 0 & 0 & 0 \\ 0 & 0 & 0 & 0 & 0 & 0 & 0 & 0 & I_{n_9} & 0 & 0 & 0 \\ 0 & I_{n_2} & 0 & 0 & 0 & 0 & 0 & 0 & 0 & 0 & 0 & 0 \\ 0 & 0 & 0 & 0 & 0 & I_{n_6} & 0 & 0 & 0 & 0 & 0 & 0 \\ 0 & 0 & 0 & 0 & 0 & 0 & 0 & 0 & 0 & I_{n_{10}} & 0 & 0 \\ 0 & 0 & I_{n_3} & 0 & 0 & 0 & 0 & 0 & 0 & 0 & 0 & 0 \\ 0 & 0 & 0 & 0 & 0 & 0 & I_{n_7} & 0 & 0 & 0 & 0 & 0 \\ 0 & 0 & 0 & 0 & 0 & 0 & 0 & 0 & 0 & 0 & I_{n_{11}} & 0 \\ 0 & 0 & 0 & I_{n_4} & 0 & 0 & 0 & 0 & 0 & 0 & 0 & 0 \\ 0 & 0 & 0 & 0 & 0 & 0 & 0 & I_{n_8} & 0 & 0 & 0 & 0 \\ 0 & 0 & 0 & 0 & 0 & 0 & 0 & 0 & 0 & 0 & 0 & I_{n_{12}} \end{bmatrix}. \quad ■$$

We now prove the following auxiliary, rather obvious, result.

Theorem 4.2 (Minc [3]). *If A is a reducible matrix in the form* (1) *with no zero rows or zero columns, then there exist increasing sequences $\alpha^1, \alpha^2, \ldots, \alpha^k, \beta^1, \beta^2, \ldots, \beta^k$, such that*

$$\alpha^i \cup \beta^i = \{1, 2, \ldots, n_i\}, \qquad \alpha^i \cap \beta^i = \varnothing, \qquad i = 1, 2, \ldots, k,$$

and

$$A_{12}[\alpha^1|\beta^2] = 0, \qquad A_{23}[\alpha^2|\beta^3] = 0, \ldots, A_{k1}[\alpha^k|\beta^1] = 0.$$

Proof. Clearly, if A is an $n \times n$ reducible matrix, then there exist sequences $\lambda = (j_1, j_2, \ldots, j_s) \in Q_{s,n}$ and $\mu = (j_{s+1}, j_{s+2}, \ldots, j_n) \in Q_{n-s,n}$, such that $\lambda \cup \mu = \{1, 2, \ldots, n\}$ and $A[\lambda|\mu] = 0$ (Problem 6). Partition λ and μ conformally with the partition of A. Let λ^i be the subsequence of λ all of whose elements lie in the interval $(\sum_{r=1}^{i-1} n_r, \sum_{r=1}^{i} n_r]$, and let μ^i be the subsequence of μ all of whose elements lie in the same interval, $i = 1, 2, \ldots, k$, where $\sum_{r=1}^{0} n_r$ is interpreted as 0. Then, clearly,

$$\lambda^i \cup \mu^i = \left\{ \sum_{r=1}^{i-1} n_r + 1, \sum_{r=1}^{i-1} n_r + 2, \ldots, \sum_{r=1}^{i} n_r \right\},$$

and (3)

$$\lambda^i \cap \mu = \lambda \cap \mu^i = \varnothing, \qquad i = 1, 2, \ldots, k, \qquad \bigcup_{i=1}^{k} \lambda^i = \lambda, \qquad \bigcup_{i=1}^{k} \mu^i = \mu.$$

Moreover, since none of the submatrices $A_{i,i+1}$ can have a zero row or a zero column, none of the sequences $\lambda^1, \lambda^2, \ldots, \lambda^k, \mu^1, \mu^2, \ldots, \mu^k$ can be empty (see Problem 7).

Now, let α^i be the sequence obtained from λ^i by subtracting $\sum_{r=1}^{i-1} n_r$ from each of its elements, and let β^i be the sequence derived from μ^i in the same manner, $i = 1, 2, \ldots, k$. Then, by (3),

$$\alpha^i \cup \beta^i = \{1, 2, \ldots, n_i\}, \qquad \alpha^i \cap \beta^i = \varnothing, \qquad i = 1, 2, \ldots, k.$$

Furthermore, if $p \in \alpha^i$ and $q \in \beta^{i+1}$, then $u = p + \sum_{r=1}^{i-1} n_r \in \lambda$ and $v = q + \sum_{r=1}^{i} n_r \in \mu$, and thus the (u, v) entry of A is zero. But the (u, v) entry of A is the (p, q) entry of $A_{i,i+1}$, and thus

$$A_{i,i+1}[\alpha^i|\beta^{i+1}] = 0, \qquad i = 1, 2, \ldots, k-1, \qquad A_{k1}[\alpha^k|\beta^1] = 0. \quad \blacksquare$$

We use the preceding result to prove the following theorem which gives necessary and sufficient conditions for a matrix in the superdiagonal block form to be irreducible.

Theorem 4.3 (Minc [4]). *Let A be a nonnegative matrix in the form (1) without zero rows or zero columns. Then A is irreducible if and only if the product $A_{12}A_{23} \cdots A_{k1}$ is irreducible.*

Proof. We prove the sufficiency of the condition by showing that if a matrix A in the form (1) is reducible but has no zero rows or zero columns, then $A_{12}A_{23} \cdots A_{k1}$ must be reducible. By Theorem 4.2, there exist nonempty index sets $\alpha^1, \alpha^2, \ldots, \alpha^k, \beta^1, \beta^2, \ldots, \beta^k$ (see Problem 7) such that $\alpha^i \cup \beta^i = \{1, 2, \ldots, n_i\}$, $\alpha^i \cap \beta^i = \varnothing$, $i = 1, 2, \ldots, k$, and

$$A_{12}[\alpha^1|\beta^2] = 0, \ A_{23}[\alpha^2|\beta^3] = 0, \ldots, A_{k1}[\alpha^k|\beta^1] = 0. \tag{4}$$

We assert that

$$(A_{12}A_{23}\ \cdots\ A_{k1})[\alpha^1|\beta^1] = 0,$$

and thus that $A_{12}A_{23}\ \cdots\ A_{k1}$ is reducible. Let $p \in \alpha^1$ and $q \in \beta^1$. The (p, q) entry of $A_{12}A_{23}\ \cdots\ A_{k1}$ is

$$\sum_{t_1, t_2, \ldots, t_{k-1}} (A_{12})_{pt_1}(A_{23})_{t_1t_2}\ \cdots\ (A_{k-1,k})_{t_{k-2}t_{k-1}}(A_{k1})_{t_{k-1}q}. \tag{5}$$

We show that every term in the sum (5) is zero. Indeed,

(i) either $(A_{12})_{pt_1} = 0$ or, by (4), $t_1 \notin \beta^2$ and therefore $t_1 \in \alpha^2$; it follows that

(ii) either $(A_{23})_{t_1t_2} = 0$ or, by (4), $t_2 \notin \beta^3$ and therefore $t_2 \in \alpha^3$; it follows that

(iii) either $(A_{34})_{t_2t_3} = 0$ or, by (4), $t_3 \notin \beta^4$ and therefore $t_3 \in \alpha^4$; and so on.

(iv) If the first $k - 1$ factors of

$$(A_{12})_{pt_1}(A_{23})_{t_1t_2}\ \cdots\ (A_{k-1,k})_{t_{k-2}t_{k-1}}(A_{k1})_{t_{k-1}q} \tag{6}$$

are nonzero, then $t_{k-1} \notin \beta^k$ and therefore $t_{k-1} \in \alpha^k$. But then $(A_{k1})_{t_{k-1}q} = 0$, since $q \in \beta^1$. Hence one of the factors in the product (6) must be zero for any choice of $t_1, t_2, \ldots, t_{k-1}$. It follows that

$$(A_{12}A_{23}\ \cdots\ A_{k1})[\alpha^1|\beta^1] = 0,$$

and thus $A_{12}A_{23}\ \cdots\ A_{k1}$ is reducible.

We prove the converse. Let A be an irreducible matrix in the form (1). Let h be the index of A, and let

$$\lambda_t = re^{i2\pi t/h}, \qquad t = 0, 1, \ldots, h-1,$$

be the h eigenvalues of A of maximum modulus r. Let x be a positive eigenvector corresponding to r. By Lemma 4.1,

$$A^k = \sum_{j=1}^{k} B_j,$$

where $B_j = A_{j,j+1}A_{j+1,j+2}\ \cdots\ A_{j-1,j}$, $j = 1, 2, \ldots, k$, and, by Lemma 4.2, each of the B_j has the same nonzero eigenvalues. If $h = km$ (see Theorem 4.1), then the eigenvalues of A^k of the maximum modulus r^k are

$$r^k e^{i2\pi t/m}, \qquad t = 0, 1, \ldots, h-1.$$

It follows that exactly $h/m = k$ of them are actually equal to r^k. Thus each of the B_j has a single maximal real eigenvalue r^k. We now partition the positive eigenvector x into blocks conformally to the partition of A:

$$x = \sum_{j=1}^{k} {}^{\cdot}\, x^j,$$

where x^j is a positive n_j-tuple, $j = 1, 2, \ldots, k$. We have

$$A^k x = \sum_{j=1}^{k} {}^{\cdot}\, B_j x^j,$$

and

$$\begin{aligned} A^k x &= r^k x \\ &= \sum_{j=1}^{k} {}^{\cdot}\, r^k x^j. \end{aligned}$$

It follows that

$$B_j x^j = r^k x^j, \qquad j = 1, 2, \ldots, k.$$

Hence each B_j has a simple maximal real eigenvalue r^k and a positive eigenvector corresponding to it. Similarly, each B_j^{T} has a positive eigenvector corresponding to r^k. Thus, by Theorem 4.7, Chapter I, each $B_j = A_{j,j+1} A_{j+1,j+2} \cdots A_{j-1,j}$ is irreducible. ∎

Corollary 4.1 (Minc [4]). *Let A be a nonnegative matrix in the form* (1) *without zero rows or zero columns. Then A is irreducible with index of imprimitivity h if and only if $A_{12}A_{23} \cdots A_{k1}$ is irreducible with index h/k. In particular, A is irreducible with index k if and only if $A_{12}A_{23} \cdots A_{k1}$ is primitive.*

For an alternative, shorter proof of the sufficiency of the condition in Theorem 4.3 see [7].

We note en passant the following simple result on the minimum number of zero eigenvalues that a matrix in the form (1) must have.

Theorem 4.4 (Minc [3]). *Let A be a nonnegative matrix in the form* (1). *Then the number of zero eigenvalues of A is at least*

$$n - kn_\mu,$$

where $n_\mu = \min(n_1, n_2, \ldots, n_k)$.

Proof. By Lemma 4.1,

$$A^k = A_{12}A_{23} \cdots A_{k1} \dotplus A_{23}A_{34} \cdots A_{12} \dotplus \cdots \dotplus A_{k1}A_{12} \cdots A_{k-1,k}.$$

The ranks of the above products cannot exceed n_μ, the least of the n_i. Hence $A_{i,i+1}A_{i+1,i+2} \cdots A_{i-1,i}$, which is n_i-square, has at least $n_i - n_\mu$ zero eigenvalues. It follows that A^k, and thus A, has at least

$$\sum_{i=1}^{k} (n_i - n_\mu) = n - kn_\mu$$

zero eigenvalues. ■

Corollary 4.2 (Minc [3]). *If A is a nonnegative nonsingular matrix in the form* (1), *then every* $A_{i,i+1}$ *is* n/k*-square.*

Of course, the result in Corollary 4.2 can be proved easily without the use of Theorem 4.4 (see Problem 9).

The next two theorems extend the results in Lemmas 4.1 and 4.2.

Theorem 4.5 (Minc [5]). *Let* $B_1, B_2, \ldots, B_s$ *and* $C_1, C_2, \ldots, C_t$ *be irreducible nonnegative matrices. The direct sums*

$$G = \sum_{i=1}^{s} {}^{\cdot} B_i$$

and

$$H = \sum_{i=1}^{t} {}^{\cdot} C_i$$

are cogredient if and only if $s = t$, *and there exists a permutation* $\sigma \in S_k$ *such that* B_i *and* $C_{\sigma(i)}$ *are cogredient for* $i = 1, 2, \ldots, k$.

Proof. The sufficiency of the conditions is quite obvious. To prove the necessity, let P be a permutation matrix such that

$$P^{\mathrm{T}}GP = H,$$

and let τ be the permutation corresponding to P, so that the (i, j) entry of G is permuted into the $(\tau(i), \tau(j))$ position of $H = P^{\mathrm{T}}GP$. For brevity, the notation $\bar{i}$ is used in place of $\tau(i)$. If $A[\mu_1, \mu_2, \ldots, \mu_a | \nu_1, \nu_2, \ldots, \nu_b]$ denotes the submatrix of A lying in rows numbered $\mu_1, \mu_2, \ldots, \mu_a$ and columns numbered $\nu_1, \nu_2, \ldots, \nu_b$, then rows $\mu_1, \mu_2, \ldots, \mu_a$ of A (and columns $\nu_1, \nu_2, \ldots, \nu_b$) are said to *intersect* the submatrix. Now, suppose that for some v, $1 \leq v \leq t$,

$$C_v = H[\bar{\alpha}_1, \bar{\alpha}_2, \ldots, \bar{\alpha}_p, \bar{\beta}_{p+1}, \bar{\beta}_{p+2}, \ldots, \bar{\beta}_q | \bar{\alpha}_1, \bar{\alpha}_2, \ldots, \bar{\alpha}_p, \bar{\beta}_{p+1}, \bar{\beta}_{p+2}, \ldots, \bar{\beta}_q],$$

and that rows and columns $\alpha_1, \alpha_2, \ldots, \alpha_p$ of G intersect block B_u, but none of rows or columns $\beta_{p+1}, \beta_{p+2}, \ldots, \beta_q$ of G intersects B_u. Now, the only nonzero entries in the rows $\alpha_1, \alpha_2, \ldots, \alpha_p$ of G are in the columns $\alpha_1, \alpha_2, \ldots, \alpha_p$. Thus

$$G[\alpha_1, \alpha_2, \ldots, \alpha_p | \beta_{p+1}, \beta_{p+2}, \ldots, \beta_q] = 0,$$

and therefore

$$H[\bar{\alpha}_1, \bar{\alpha}_2, \ldots, \bar{\alpha}_p | \bar{\beta}_{p+1}, \bar{\beta}_{p+2}, \ldots, \bar{\beta}_q] = 0.$$

However, this would imply that C_v is reducible. Hence the supposition is impossible, and each of the C_j can be intersected only by rows and columns which correspond to rows and columns that intersect a single B_i. Since $\Sigma_{i=1}^{\cdot t} C_i$ and $\Sigma_{i=1}^{\cdot s} B_i$ are cogredient, the result follows. ■

Theorem 4.6 (Minc [5]). *If A is an irreducible nonnegative matrix and if A^k is cogredient to a direct sum of irreducible matrices $C_1, C_2, \ldots, C_k$, then k divides the index of imprimitivity of A, and all the C_i have the same nonzero eigenvalues.*

Proof. Let P be a permutation matrix such that $PA^kP^{\mathrm{T}} = \Sigma_{t=1}^{\cdot k} C_t$. Let r be the maximal eigenvalue and x a positive maximal eigenvector of A. Then

$$\left(\sum_{t=1}^{k}{}^{\cdot}\, C_t\right) Px = PA^kP^{\mathrm{T}}Px = PA^kx = r^kPx.$$

It follows that r^k is an eigenvalue (clearly maximal) of each of the C_t. Since the C_t are irreducible, the eigenvalue r^k is simple, and therefore A^k has exactly k eigenvalues equal to r^k. But Theorem 1.1 implies that there are $d = \text{g.c.d.}(h, k)$ such eigenvalues. Hence $d = k$, and thus k divides h.

It now follows from Theorem 4.1, in conjunction with Lemmas 4.1 and 4.2 and Theorem 4.3, that A^k is cogredient to a matrix of the form

$$\sum_{t=1}^{k}{}^{\cdot}\, B_t,$$

where the B_t are irreducible and all the B_t have the same nonzero eigenvalues. But then $\Sigma_{t=1}^{\cdot k} B_t$ and $\Sigma_{t=1}^{\cdot k} C_t$ are cogredient, and all the B_t and all the C_t are irreducible. Thus, by Theorem 4.5, the $B_1, B_2, \ldots, B_k$ are cogredient to the $C_1, C_2, \ldots, C_k$, in some order, and the result follows. ■

We use the preceding theorems to extend a result of Mirsky [6] and apply it to imprimitive matrices. We shall require the following auxiliary result on matrices in the form (1).

Theorem 4.7 (Minc [5]). *Let A be a complex $n \times n$ matrix in the form* (1), *and let*

$$\lambda^n + \sum_t b_t \lambda^{m_t},$$

where the coefficients b_t are nonzero, be the characteristic polynomial of A. Then k divides $n - m_t$ for all t.

Proof. Let $p(\lambda, M)$ denote the characteristic polynomial of M. Suppose that A is in the form (1), where the block $A_{s,s+1}$ is $n_s \times n_{s+1}$, $s = 1, 2, \ldots, k-1$, and A_{k1} is $n_k \times n_1$, and let

$$D = \sum_{s=1}^{k} {}^{\cdot}\, \theta^s I_{n_s},$$

where $\theta = \exp(2\pi i/k)$. Then

$$D^{-1}AD = \theta A,$$

and therefore

$$D^{-1}(\theta\lambda I_n - A)D = \theta(\lambda I_n - A),$$

so that

$$p(\theta\lambda, A) = \theta^n p(\lambda, A).$$

Hence

$$\theta^n\lambda^n + \sum_t b_t \theta^{m_t}\lambda^{m_t} = \theta^n\lambda^n + \sum_t b_t \theta^n \lambda^{m_t},$$

that is,

$$\theta^{m_t} = \theta^n,$$

for all t. Thus

$$\exp(2\pi i(n - m_t)/k) = 1,$$

for all t. The result follows. ■

Our next result is a generalization of a theorem of Mirsky [6]. The proof is similar to Mirsky's proof in [6].

Theorem 4.8 (Minc [5]). *Let A be an $n \times n$ complex matrix in the form* (1), *and suppose that $\omega_1, \omega_2, \ldots, \omega_m$ are the nonzero eigenvalues of the product $A_{12}A_{23} \cdots A_{k1}$. Then the spectrum of A consists of $n - km$ zeros and the km kth roots of the numbers $\omega_1, \omega_2, \ldots, \omega_m$.*

Proof. By Lemma 4.2, the spectrum of A^k consists of the numbers $\omega_1, \omega_2, \ldots, \omega_m$, each counted k times, and $n - km$ zeros. Thus

$$p(\lambda, A^k) = \lambda^{n-km} \prod_{j=1}^{m} (\lambda - \omega_j)^k, \tag{7}$$

and therefore

$$p(\lambda, A) = \lambda^{n-km} \varphi(\lambda),$$

where

$$\varphi(\lambda) = \sum_{t=0}^{km} c_t \lambda^t.$$

By Theorem 4.7, coefficient c_t must vanish, unless k divides $n - (n - km + t) = km - t$. It follows that $c_t = 0$ whenever k does not divide t. In other words, $\varphi(\lambda)$ is a polynomial in λ^k:

$$\varphi(\lambda) = \prod_{t=1}^{m} (\lambda^k - \zeta_t),$$

for some numbers $\zeta_1, \zeta_2, \ldots, \zeta_m$. Hence

$$\begin{aligned} p(\lambda, A) &= \lambda^{n-km} \prod_{t=1}^{m} (\lambda^k - \zeta_t) \\ &= \lambda^{n-km} \prod_{\substack{1 \le t \le m \\ 1 \le j \le k}} (\lambda - \zeta_t^{1/k} \theta^j), \end{aligned} \tag{8}$$

where $\theta = \exp(2\pi i/k)$ and $\zeta_t^{1/k}$ denotes any fixed kth root of ζ_t. Therefore the characteristic polynomial of A^k is

$$p(\lambda, A^k) = \lambda^{n-km} \prod_{t=1}^{m} (\lambda - \zeta_t)^k. \tag{9}$$

Comparing (7) and (9) it can be concluded that the numbers $\zeta_1, \zeta_2, \ldots, \zeta_m$ are the same as the numbers $\omega_1, \omega_2, \ldots, \omega_m$, in some order. Thus the characteristic polynomial (8) of A reads

$$p(\lambda, A) = \lambda^{n-km} \prod_{\substack{1 \le t \le m \\ 1 \le j \le k}} (\lambda - \omega_t^{1/k}\theta^j),$$

and the theorem is established. ■

We now exploit Theorem 4.8 to gain information about the spectra of imprimitive matrices.

Theorem 4.9 (Minc [5]). *Let A be an irreducible $n \times n$ matrix, and suppose that A^k is cogredient to a direct sum of irreducible matrices $C_1, C_2, \ldots, C_k$. If the nonzero eigenvalues of C_1 are $\omega_1, \omega_2, \ldots, \omega_m$, then the spectrum of A consists of $n - km$ zeros and the km kth roots of $\omega_1, \omega_2, \ldots, \omega_m$.*

Proof. By Theorem 4.6, k divides the index of A, and thus, by Theorem 4.1, the matrix A is cogredient to a matrix in the form (1) with blocks $A_{12}, A_{23}, \ldots, A_{k1}$ in the superdiagonal. Then A^k is cogredient to

$$\sum_{t=1}^{k}{}^{\cdot} B_t,$$

where $B_t = A_{t,t+1}A_{t+1,t+2} \cdots A_{t-1,t}$, $t = 1, 2, \ldots, k$ (subscripts reduced modulo k), and all the B_t have the same nonzero eigenvalues. Hence, by Theorems 4.5 and 4.6, the matrices B_1 and C_1 have the same nonzero eigenvalues. The result now follows by virtue of Theorem 4.8. ■

PROBLEMS

1 Prove that $\lim_{m\to\infty} A^m = 0$ if and only if the spectrum of A lies in the interior of the unit circle.

2 Find pairs of imprimitive 3×3 matrices A and B satisfying the following conditions, or show that such matrices do not exist:

(a) AB is primitive (is BA then necessarily primitive?);

(b) AB is imprimitive;

(c) AB is reducible.

3 Find pairs of reducible 3×3 matrices A and B satisfying the following conditions:

(a) AB is positive;

(b) AB is primitive but not positive;

(c) AB is imprimitive.

4 Find an imprimitive 3×3 matrix A and a reducible 3×3 matrix B:

(a) such that AB is primitive;

(b) such that AB is reducible.

5 Let $A_{j,j+1}$ be an $n_j \times n_{j+1}$ complex matrix for $j = 1, 2, \ldots, k$ (subscripts reduced modulo k), and let $B_j = A_{j,j+1}A_{j+1,j+2} \cdots A_{j-1,j}$, $j = 1, 2, \ldots, k$. Prove that all the B_j have the same nonzero eigenvalues. (*Hint*: First prove the result for two matrices one of which is nonsingular. Extend this result to two singular matrices, and then to rectangular matrices. Finally, generalize it to k matrices.)

6 Prove in detail that an $n \times n$ nonnegative matrix A is reducible, if and only if there exist disjoint sequences λ and μ such that $\lambda \cup \mu = \{1, 2, \ldots, n\}$ and $A[\lambda|\mu] = 0$.

7 Show that if one of the sequences $\alpha^1, \alpha^2, \ldots, \alpha^k, \beta^1, \beta^2, \ldots, \beta^k$ in Theorem 4.2 were empty, then the matrix A would have a zero row or a zero column.

8 Let A be an irreducible matrix with index of imprimitivity h. Show that A^k is irreducible if and only if h and k are relatively prime.

9 Prove Corollary 4.2 by considering the dimension of the row space or the column space of A.

10 Is the converse of the result in Corollary 4.2 true?

11 Let Q and A be the matrices in Example 4.1. Find a permutation matrix R such that $RQAQ^{\mathrm{T}}R^{\mathrm{T}}$ is in a superdiagonal block form with $k = 3$.

12 Let P be the $n \times n$ permutation matrix with 1's in the positions $(i, i+1)$, $i = 1, 2, \ldots, n-1$, and $(n, 1)$. Use Theorem 4.8 to determine the spectrum of P.

13 Let P be the permutation matrix defined in Problem 12, and let E_{13} be the $n \times n$ matrix with 1 in the $(1, 3)$ position and 0's elsewhere. Show that $P + E_{13}$ is primitive.

14 Show that an $n \times n$ permutation matrix is irreducible if and only if it is cogredient to the permutation matrix in Problem 12.

15 Show that a symmetric imprimitive matrix must have index 2.

16 Let A be an irreducible nonsingular $n \times n$ matrix with index h. How many real eigenvalues at most can A have?

17 Let A be a matrix in the form (1), Section 3.4, where the blocks $A_{12}, A_{23}, \ldots, A_{k1}$ are square. Let p be the number of zero eigenvalues of A. Show that $p \equiv 0 \pmod{k}$.

18 For each of the following matrices, either find an irreducible matrix whose square is equal to the given matrix or show that such a matrix cannot exist:

$$A = \begin{bmatrix} 1 & 0 & 0 \\ 0 & 1 & 1 \\ 0 & 1 & 1 \end{bmatrix},$$

$$B = \begin{bmatrix} 1 & 1 & 0 \\ 1 & 1 & 0 \\ 0 & 0 & 2 \end{bmatrix},$$

$$C = \begin{bmatrix} 1 & 1 & 0 & 0 \\ 1 & 1 & 0 & 0 \\ 0 & 0 & 1 & 1 \\ 0 & 0 & 1 & 1 \end{bmatrix},$$

$$D = \begin{bmatrix} 1 & 1 & 0 & 0 \\ 1 & 1 & 0 & 0 \\ 0 & 0 & 2 & 0 \\ 0 & 0 & 0 & 0 \end{bmatrix},$$

$$E = \begin{bmatrix} 0 & 1 & 0 & 0 \\ 1 & 0 & 0 & 0 \\ 0 & 0 & 0 & 1 \\ 0 & 0 & 1 & 0 \end{bmatrix}.$$

REFERENCES

1. G. Frobenius, Über Matrizen aus nicht negativen Elementen, *S.-B. K. Preuss. Akad. Wiss. Berlin* (1912), 456–477.
2. M. Marcus and H. Minc, On two theorems of Frobenius, *Pacific J. Math.* **60** (1975), 149–151.
3. H. Minc, Irreducible matrices, *Linear and Multilinear Algebra* **1** (1974), 337–342.
4. H. Minc, The structure of irreducible matrices, *Linear and Multilinear Algebra* **2** (1974), 85–90.
5. H. Minc, Spectra of irreducible matrices, *Proc. Edinburgh Math. Soc.* **19** (1975), 231–236.
6. L. Mirsky, An inequality for characteristic roots and singular values of complex matrices, *Monatsh. Math.* **70** (1966), 357–359.
7. N. J. Pullman, A note on a theorem of Minc on irreducible nonnegative matrices, *Linear and Multilinear Algebra* **4** (1975), 335–336.
8. J. J. Sylvester, On the equation to the secular inequalities in the planetary theory, *Philos. Mag.* **16**(5) (1883), 267–269.
9. H. Wielandt, Unzerlegbare nicht negative Matrizen, *Math. Z.* **52** (1950), 642–648.

IV

Structural Properties of Nonnegative Matrices

4.1. (0, 1)-MATRICES. PERMANENTS

We now turn our attention to properties of nonnegative matrices which depend on their zero pattern only. From this point of view, matrices are essentially rectangular arrays with entries of two kinds. However, in most combinatorial applications it is convenient to represent these entries by 0 and 1, particularly if the permanent function, or some other combinatorial matrix function, is used for enumerative purposes. A matrix each of whose entries is either 0 or 1, is called a $(0,1)$-*matrix*.

Let $S_1, S_2, \ldots, S_m$ be subsets, not necessarily distinct, of an n-set $S = \{x_1, x_2, \ldots, x_n\}$. Let $A = (a_{ij})$ be the $m \times n$ $(0,1)$-matrix with $a_{ij} = 1$ if $x_j \in S_i$, and $a_{ij} = 0$ if $x_j \notin S_i$. The matrix A is called the *incidence matrix* of the subset configuration $S_1, S_2, \ldots, S_m$. Once the x_j's and the S_i's are ordered, the incidence matrix is uniquely determined by the configuration, and vice versa.

The definition of incidence matrices can also be specialized to representations of relations, functions, graphs, set intersections, etc.

Definition 1.1. Let $S_1, S_2, \ldots, S_m$ be subsets of an n-set S. A sequence $(s_1, s_2, \ldots, s_m)$ of m distinct elements of S is said to form a *system of distinct representatives* (abbreviated to *SDR*) for the configuration $S_1, S_2, \ldots, S_m$, if $s_i \in S_i$, $i = 1, 2, \ldots, m$.

A configuration of subsets may or may not have an SDR. It is of considerable interest in combinatorics to determine whether a given configuration has any SDRs, and if so, how many.

Example 1.1. (a) The configuration of subsets $X_1 = \{x_1, x_3\}$, $X_2 = \{x_2, x_3, x_4\}$, $X_3 = \{x_1, x_2\}$, $X_4 = \{x_1, x_3, x_4\}$ of the 4-set $X = \{x_1, x_2, x_3, x_4\}$ has four SDRs: (x_1, x_3, x_2, x_4), (x_1, x_4, x_2, x_3), (x_3, x_2, x_1, x_4), (x_3, x_4, x_2, x_1).

(b) The configuration of four subsets, $S_1 = S_3 = S_4 = \{s_2, s_4\}$, $S_2 = S$, of a 5-set $S = \{s_1, s_2, s_3, s_4, s_5\}$ has no SDRs. For, $S_1 \cup S_3 \cup S_4$ contains only two elements, and therefore the subsets S_1, S_3, and S_4 cannot be represented by three distinct representatives. ■

The problems of determining the existence of SDRs of a given configuration of subsets and of evaluating the number of SDRs can be analyzed conveniently using incidence matrices.

Let $A = (a_{ij})$ be the incidence matrix for subsets $S_1, S_2, \dots, S_m$ of an n-set $\{x_1, x_2, \dots, x_n\}$. If the configuration has an SDR, then clearly $m \leq n$ and there exists a one–one function $\sigma\colon \{1, 2, \dots, m\} \to \{1, 2, \dots, n\}$ such that

$$x_{\sigma(i)} \in S_i, \qquad i = 1, 2, \dots, m.$$

It follows from the definition of incidence matrices that

$$a_{i\sigma(i)} = 1, \qquad i = 1, 2, \dots, m.$$

Hence the configuration has an SDR if and only if there exists a one–one function σ such that

$$\prod_{i=1}^{m} a_{i\sigma(i)} = 1. \tag{1}$$

The number of SDRs is equal to the number of distinct one–one functions σ for which (1) holds. It is therefore equal to

$$\sum_{\sigma} \prod_{i=1}^{m} a_{i\sigma(i)}, \tag{2}$$

where the summation is over all one–one functions from $\{1, 2, \dots, m\}$ to $\{1, 2, \dots, n\}$.

Definition 1.2. Let $A = (a_{ij})$ be an $m \times n$ matrix with complex or real entries, $m \leq n$. The *permanent* of A is defined by

$$\operatorname{Per}(A) = \sum_{\sigma} \prod_{i=1}^{m} a_{i\sigma(i)}, \tag{3}$$

where the summation extends over all one–one functions σ as in (2). The special case $m = n$ is of particular importance; in this case we write per(A) instead of Per(A). Thus, if $A = (a_{ij})$ is n-square,

$$\operatorname{per}(A) = \sum_{\sigma \in S_n} \prod_{i=1}^{n} a_{i\sigma(i)}. \tag{4}$$

Our conclusions about the existence of an SDR for a configuration of subsets and the number of such SDRs can be restated in terms of permanents as follows. *A configuration has an SDR if and only if its incidence matrix has a positive permanent. The number of SDRs for a configuration equals the permanent of the incidence matrix of the configuration.*

The similarity between the definition of the determinant function and that of the permanent function on square matrices is quite apparent. In fact, permanents do possess some properties analogous to those of determinants.

Theorem 1.1. *Let A be an $m \times n$ matrix, $m \leq n$.*

(a) *The permanent of A is a multilinear function of rows of A.*

(b) *If $m = n$, then* $\operatorname{per}(A^{\mathrm{T}}) = \operatorname{per}(A)$.

(c) *If P and Q are $m \times m$ and $n \times n$ permutation matrices, respectively, then*

$$\operatorname{per}(PAQ) = \operatorname{per}(A).$$

(d) *If D and G are $m \times m$ and $n \times n$ diagonal matrices, respectively, then*

$$\operatorname{per}(DAG) = \operatorname{per}(D)\operatorname{per}(A)\operatorname{per}(G).$$

All these properties are immediate consequences of the definition of permanents.

The next theorem is an analogue of the Laplace expansion theorem for determinants.

Theorem 1.2. *Let $A = (a_{ij})$ be an $m \times n$ matrix, $m \leq n$, and let α be a sequence in $Q_{r,m}$. Then, for $r < m$,*

$$\operatorname{Per}(A) = \sum_{\omega \in Q_{r,n}} \operatorname{per}(A[\alpha|\omega]) \cdot \operatorname{Per}(A(\alpha|\omega)). \tag{5}$$

In particular, for any i, $1 \leq i \leq m$,

$$\operatorname{Per}(A) = \sum_{t=1}^{n} a_{it} \operatorname{Per}(A(i|t)). \tag{6}$$

In the case $m = n$, analogous formulas hold for expansions by columns.

Proof. Consider the entries of A as indeterminates. For a particular $\omega \in Q_{r,n}$ the permanent of $A[\alpha|\omega]$ is a sum of $r!$ diagonal products, and that of $A(\alpha|\omega)$ is a sum of $\binom{n-r}{m-r}(m-r)!$ diagonal products. The product of a diagonal product of $A[\alpha|\omega]$ by a diagonal product of $A(\alpha|\omega)$ is a diagonal product of A. Thus, for a fixed ω, $\operatorname{per}(A[\alpha|\omega]) \cdot \operatorname{Per}(A(\alpha|\omega))$ is a sum of $r!\binom{n-r}{m-r}(m-r)!$ distinct diagonal products of A. Furthermore, for different sequences ω, different diagonal products are obtained. Now, there are $\binom{n}{r}$

sequences in $Q_{r,n}$, and therefore the right-hand side of (5) is the sum of

$$\binom{n}{r} r! \binom{n-r}{m-r} (m-r)! = \binom{n}{m} m!$$

such diagonal products, that is, the sum of all the diagonal products of A, and thus is equal to $\text{Per}(A)$. ∎

For a comprehensive study of permanents see the monograph [24].

4.2. THE FROBENIUS–KÖNIG THEOREM

The fundamental result on zero patterns of matrices is the so-called Frobenius–König theorem. It was first obtained by Frobenius [8]. In 1915 König [15] gave an elementary proof of the theorem, using graphs. In 1917 the theorem was again re-proved by Frobenius [9] using an elementary method. An acrimonious controversy developed between Frobenius and König about their respective contributions to this result (see [16]). We do not intend to adjudicate in this matter, and shall refer to the result in question (Theorem 2.1 below) as the *Frobenius–König theorem*, by which name it is generally known.

The Frobenius–König theorem states that a necessary and sufficient condition for every "term in the expansion of the determinant of an $n \times n$ matrix" to be zero, is that the matrix contain an $s \times t$ zero submatrix with $s + t = n + 1$. We restate the theorem in terms of permanents and extend it to rectangular matrices.

Theorem 2.1. *The permanent of a nonnegative $m \times n$ matrix, $m \leq n$, vanishes if and only if the matrix contains an $s \times t$ zero submatrix with $s + t = n + 1$.*

Proof. Let A be an $m \times n$ matrix, $m \leq n$, and suppose that $A[\alpha|\beta] = 0$, $\alpha \in Q_{s,m}$, $\beta \in Q_{t,n}$, and $s + t = n + 1$. Then the submatrix $A[\alpha|1, 2, \ldots, n]$ contains at most $n - t = s - 1$ nonzero columns, and therefore every $s \times s$ submatrix $A[\alpha|\omega]$, $\omega \in Q_{s,n}$, has a zero column. In other words, $\text{per}(A[\alpha|\omega]) = 0$ for every $\omega \in Q_{s,n}$. If $s = m$, the result is obvious. If $s < m$, it follows, by Theorem 1.2, that

$$\text{Per}(A) = \sum_{\omega \in Q_{s,n}} \text{per}(A[\alpha|\omega]) \cdot \text{Per}(A(\alpha|\omega)) = 0.$$

Conversely, suppose that $A = (a_{ij})$ is an $m \times n$ matrix, $m \leq n$, and $\text{Per}(A) = 0$. We use induction on m. If $m = 1$, then A must be a zero matrix. Assume that $m > 1$, and that the theorem holds for all matrices with less than m rows for which the permanent is defined. If $A = 0$, there is nothing to prove. Otherwise, A contains a nonzero entry a_{hk}. But then $\text{Per}(A(h|k)) = 0$, since $0 = \text{Per}(A) \geq a_{hk}\text{Per}(A(h|k))$. By the induction hypothesis, the submatrix $A(h|k)$ contains a $p \times q$ zero submatrix with $p + q = n$. Let P and Q be permutation matrices such that

$$PAQ = \begin{bmatrix} X & Y \\ 0 & Z \end{bmatrix},$$

where X is $(m - p) \times q$ and Z is $p \times p$. Clearly, $m - p \leq q$ and therefore

$$0 = \text{Per}(A) = \text{Per}(PAQ) = \text{Per}(X) \cdot \text{per}(Z),$$

and therefore either $\text{Per}(X) = 0$ or $\text{per}(Z) = 0$. If $\text{Per}(X) = 0$, then using the induction hypothesis again, we can conclude that X contains a $u \times v$ zero submatrix $X[i_1, i_2, \ldots, i_u | j_1, j_2, \ldots, j_v]$ with $u + v = q + 1$. But then PAQ, and thus A, contains a $(u + p) \times v$ zero submatrix, namely, $(PAQ)[i_1, i_2, \ldots, i_u, m - p + 1, m - p + 2, \ldots, m | j_1, j_2, \ldots, j_v]$, with

$$\begin{aligned}(u + p) + v &= p + (u + v) \\ &= p + q + 1 \\ &= n + 1.\end{aligned}$$

If $\text{per}(Z) = 0$, the proof is similar. ■

Example 2.1 ("Dance Problem"). In a group of n boys and n girls every boy has been introduced to k girls, and every girl has been introduced to k boys. Show that it is possible to arrange a dance in n pairs in such a way that every girl dances only with a boy to whom she was previously introduced.

Let A be the $n \times n$ matrix whose (i, j) entry is 1 if the ith boy has been introduced to the jth girl, and 0 if he has not. Then every row sum and every column sum is k. We have to prove that $\text{per}(A) > 0$, that is, there exists a positive diagonal of A, since such a diagonal would determine a permissible arrangement in pairs for the dance. Suppose that this is not the case, namely, that $\text{per}(A) = 0$. Then, by the Frobenius–König theorem, the matrix A contains a $p \times q$ zero submatrix where $p + q = n + 1$, and thus there exist permutation matrices P and Q such that

$$\text{PAQ} = \begin{bmatrix} X & Y \\ 0 & Z \end{bmatrix},$$

where X is $(n - p) \times q$ and Z is $p \times (n - q)$. Let $\sigma(M)$ denote the sum of all entries of matrix M. Then $\sigma(PAQ) = \sigma(A) = nk$. Also, $\sigma(X) = qk$ and $\sigma(Z) = pk$, since X contains all nonzero entries in the first q columns and Z

contains all nonzero entries in the last p rows of PAQ. Hence

$$\begin{aligned} nk &= \sigma(PAQ) \\ &\geq \sigma(X) + \sigma(Z) \\ &= qk + pk \\ &= (p+q)k \\ &= (n+1)k, \end{aligned}$$

which is a contradiction. We can therefore conclude that the permanent of A is positive, and it follows that the dance can be arranged in the prescribed way. ■

Theorem 2.1 can be generalized as follows.

Theorem 2.2 (König [16]). *A necessary and sufficient condition that every diagonal of an $m \times n$ matrix, $m \leq n$, contain at least k zeros is that the matrix contain an $s \times t$ zero submatrix with $s + t = n + k$.*

Proof (Aharoni [1]). Augment A to an $m \times (n + k - 1)$ matrix $B = [A : J]$ so that the first n columns of B form the matrix A and the remaining columns form an $m \times (k-1)$ matrix J all of whose entries are nonzero. Suppose that every diagonal of A contains at least k zeros. Then every diagonal of B contains a zero. For, at least $m - (k - 1)$ entries of each diagonal of B must lie in A, and these $m - (k - 1)$ entries together with $k - 1$ additional entries in A form a diagonal of B which contains at least k zeros. Hence, by Theorem 2.1, the matrix B contains an $s \times t$ zero submatrix with $s + t = (n + k - 1) + 1 = n + k$. Clearly, it must lie in A.

Conversely, if A contains an $s \times t$ zero submatrix with $s + t = (n + k - 1) + 1 = n + k$, then, by Theorem 2.1, every diagonal of B contains at least one zero. We assert that this implies that every diagonal of A contains at least k zeros. For, if a diagonal of A contained t zero entries, $t \leq k - 1$, then the $m - t$ nonzero entries of this diagonal together with appropriate t entries in J (which are all nonzero) would form a diagonal of B without any zeros. ■

Example 2.2. Find necessary and sufficient conditions for every diagonal in an $m \times n$ matrix, $m \leq n$, to contain exactly k zeros.

Let A be an $m \times n$ matrix, $m \leq n$, and suppose that every diagonal of A consists of k zeros and $m - k$ nonzero entries. By Theorem 2.2, the matrix A must contain both an $s \times t$ zero submatrix with $s + t = k + n$, and a $p \times q$ submatrix all of whose entries are nonzero, with $p + q = n + (m - k)$. Since the two submatrices cannot overlap, we must have either $s + p \leq m$ or $t + q \leq n$. But

$$(s + p) + (t + q) = 2n + m,$$

and thus if $s + p \le m$, then

$$m + t + q \ge 2n + m,$$

that is,

$$t + q \ge 2n,$$

which implies that $q = t = n$ and therefore $s = k$ and $p = m - k$. It follows that A consists of k zero rows and $m - k$ rows without zero entries. If $t + q \le n$, then we obtain

$$s + p + n \ge 2n + m,$$

that is,

$$s + p \ge n + m,$$

which is impossible unless $m = n$. In this case $s = p = n$ and $t = k$ and $q = n - k$, and the matrix A consists of k zero columns and $n - k$ columns without zero entries.

The above conditions are obviously sufficient. ■

We define now an important concept in combinatorics. It usually appears in connection with (0, 1)-matrices, but it is preferable here to define it for nonnegative matrices.

Definition 2.1. Let A be an $m \times n$ nonnegative matrix. The *term rank* of A is the maximal number of positive entries in A no two of which lie on the same line. In other words, the term rank of A is the order of the largest positive subpermanent of A.

Theorem 2.3 (König-Egerváry [16]). *The minimal number of lines in an $m \times n$ nonnegative matrix A that contain all the positive entries in A is equal to the term rank of A.*

Proof. Let r denote the term rank of A. If w lines contain all the positive entries in A, then clearly $w \ge r$. Suppose that u rows and v columns contain all the positive entries of A, $u + v = w$, where w is minimal. We can assume without loss of generality that these are the first u rows and the first v columns of A, that is,

$$A = \begin{bmatrix} B & C \\ D & 0 \end{bmatrix},$$

where B is $u \times v$. Clearly, $w \le \min(m, n)$ and thus $u \le n - v$. Now, the term rank of C is u. Otherwise, every diagonal of C would contain a zero, and, by Theorem 2.1, the matrix C would contain a $p \times q$ zero submatrix with

$p + q = 1 + n - v$. But then A would contain an $(m - u + p) \times q$ zero submatrix. It follows that all the positive entries of A would be contained in $u - p$ of its rows and $n - q$ of its columns. But

$$\begin{aligned}(u - p) + (n - q) &= u + n - 1 - n + v \\ &= u + v - 1 \\ &= w - 1,\end{aligned}$$

and w could not have been minimal. For similar reasons the term rank of D is v. Now, the term rank of A must be at least as large as the sum of those of C and D. Therefore

$$\begin{aligned}r &\geq u + v \\ &= w.\end{aligned}$$

This, together with the previously obtained inequality $r \leq w$, implies that

$$r = w. \quad \blacksquare$$

Example 2.3. The term rank of the matrix

$$A = \begin{bmatrix} 1 & 0 & 1 & 0 \\ 0 & 0 & 1 & 0 \\ 0 & 1 & 0 & 1 \\ 1 & 0 & 1 & 0 \end{bmatrix}$$

is 3, since $\text{per}(A(1|4)) > 0$ and $\text{per}(A) = 0$. We observe that the third row, the first column, and the third column contain all the positive entries of A. ■

4.3. NONNEGATIVE MATRICES AND GRAPHS

Many of the properties of nonnegative matrices, such as irreducibility, primitivity, the Frobenius form of an irreducible matrix, and its index of imprimitivity, depend on its zero pattern only. These properties can often be studied conveniently by replacing the matrix by a $(0, 1)$-matrix with the same zero pattern. With each such $(0, 1)$-matrix we can in turn associate an essentially unique directed graph (see the definition below). In this section we show how some properties of nonnegative matrices can be deduced from relevant properties of the associated directed graph. We shall not discuss here the reverse (and perhaps the more important) problem of determining structural properties of directed graphs from algebraic properties of associated matrices.

We start with a profusion of definitions, an unavoidable prerequisite of any study of graphs.

Definition 3.1. Let V be a nonempty n-set whose elements may be conveniently labelled $1, 2, \ldots, n$, and let E be a binary relation on V, that is, a set of ordered pairs of elements of V. The pair $D = (V, E)$ is called a *directed graph*, the elements of V are its *vertices*, and the elements of E are the *arcs* of D. The arc (i, j) is said to *join* vertex i to vertex j. A *subgraph* of D is a directed graph all of whose vertices and arcs belong to D. A *spanning* subgraph of D is a subgraph of D containing all the vertices of D.

It is convenient to represent a directed graph D by means of a diagram in which the vertices of D are represented by points and its arcs by directed lines joining the appropriate points. It is customary to refer to the diagram as the directed graph D.

Example 3.1. Let $V = \{1, 2, 3, 4, 5\}$ and $E = \{(1, 1), (1, 3), (2, 5), (3, 1), (3, 4), (4, 1), (4, 2), (4, 3), (5, 4)\}$. Then the graph $D = (V, E)$ may be represented by the diagram

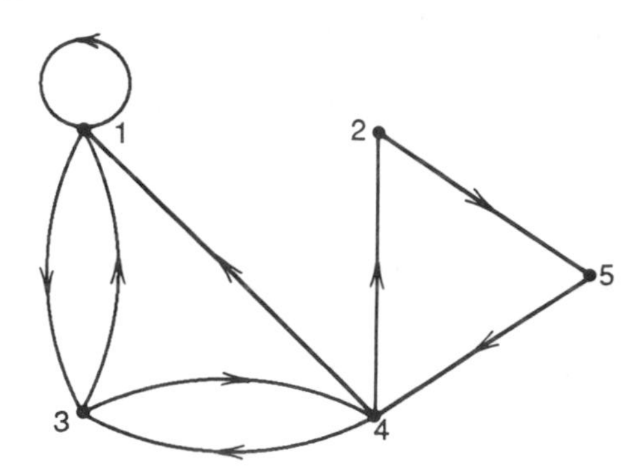

Of course, the following diagram represents the same graph

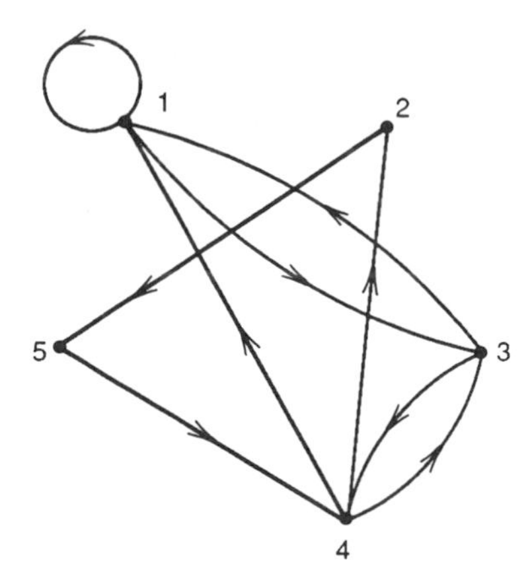

since the corresponding points (vertices) in the two diagrams are simultaneously either joined by directed lines (arcs) or are not thus connected. Note that in the second diagram some lines intersect in points which are not vertices of the graph; these spurious intersections are not part of the graph. ■

A sequence of arcs $(i, t_1), (t_1, t_2), (t_2, t_3), \ldots, (t_{m-2}, t_{m-1}), (t_{m-1}, j)$ in D is called a *path connecting i to j*. The *length* of the path is defined to be the number m of arcs in the sequence. A path of length m connecting vertex i to itself is called a *cycle of length m*. If each vertex in a cycle appears exactly once as the first vertex of an arc, the cycle is called a *circuit*. A cycle of length 1 is a *loop*. A spanning circuit is called a *Hamiltonian* circuit.

Definition 3.2. (a) The *adjacency* matrix $A(D)$ of a directed graph D with n vertices is the (0, 1)-matrix whose (i, j) entry is 1 if and only if (i, j) is an arc of D.

(b) A directed graph $D(X)$ is said to be *associated* with a nonnegative matrix X, if the adjacency matrix of $D(X)$ has the same zero pattern as X.

For example, the directed graph in Example 3.1 is associated with matrix

$$\begin{bmatrix} \frac{1}{2} & 0 & \frac{1}{2} & 0 & 0 \\ 0 & 0 & 0 & 0 & 1 \\ \frac{2}{3} & 0 & 0 & \frac{1}{3} & 0 \\ \frac{1}{4} & \frac{1}{2} & \frac{1}{4} & 0 & 0 \\ 0 & 0 & 0 & 1 & 0 \end{bmatrix}$$

(and with every other nonnegative matrix with the same zero pattern).

Definition 3.3. A directed graph D is said to be *strongly connected* if for any ordered pair of distinct vertices, i and j, there is a path in D connecting i to j.

Example 3.2. The directed graph in Example 3.1 is strongly connected whereas the graph

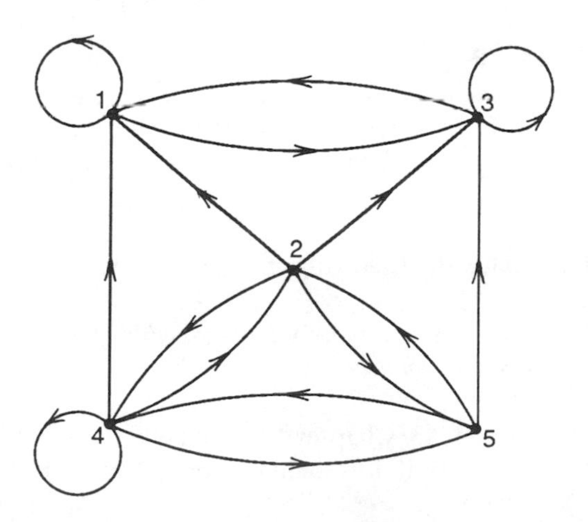

is not strongly connected: It does not contain any paths connecting 1 to 2. The adjacency matrix of this graph is

$$\begin{bmatrix} 1 & 0 & 1 & 0 & 0 \\ 1 & 0 & 1 & 1 & 1 \\ 1 & 0 & 1 & 0 & 0 \\ 1 & 1 & 0 & 1 & 1 \\ 0 & 1 & 1 & 1 & 0 \end{bmatrix},$$

which clearly is reducible. ■

In Examples 3.1 and 3.2 the associated directed graph of an irreducible matrix is strongly connected whereas that of a reducible matrix is not. We show that this correspondence is true for all nonnegative matrices.

Recall that if $A = (a_{ij})$ is a square matrix, then $a_{ij}^{(k)}$ denotes the (i, j) entry in A^k.

Theorem 3.1. *If $A = (a_{ij})$ is a $(0,1)$-matrix and $D(A)$ is the associated directed graph with vertices $1, 2, \ldots, n$, then the number of distinct paths of length k connecting vertex i to vertex j is equal to $a_{ij}^{(k)}$.*

Proof. On the one hand, we have

$$a_{ij}^{(k)} = \sum_{t_1, t_2, \ldots, t_{k-1}} a_{it_1} a_{t_1 t_2} a_{t_2 t_3} \cdots a_{t_{k-2} t_{k-1}} a_{t_{k-1} j},$$

where $t_1, t_2, \ldots, t_{k-1}$ run independently over all integers between 1 and n. On the other hand, a path $(i, t_1), (t_1, t_2), (t_2, t_3), \ldots, (t_{k-2}, t_{k-1}), (t_{k-1}, j)$ connects i to j in $D(A)$ if and only if $a_{1t_1} = a_{t_1 t_2} = \cdots = a_{t_{k-2} t_{k-1}} = a_{t_{k-1} j} = 1$, that is, if and only if

$$a_{it_1} a_{t_1 t_2} a_{t_2 t_3} \cdots a_{t_{k-2} t_{k-1}} a_{t_{k-1} j} = 1.$$

The result follows. ■

Corollary 3.1. *If $A = (a_{ij})$ is a nonnegative matrix, then the associated directed graph has a path of length k connecting vertex i to vertex j if and only if $a_{ij}^{(k)} > 0$.*

The corollary yields the following important theorem.

Theorem 3.2. *A nonnegative matrix is irreducible if and only if the associated directed graph is strongly connected.*

Proof. By Theorem 2.3, Chapter I, a nonnegative $n \times n$ matrix $A = (a_{ij})$ is irreducible if and only if for each i and j, $1 \leq i$, $j \leq n$, there exists an integer k such that $a_{ij}^{(k)} > 0$. By Corollary 3.1, this condition is satisfied if and only if the associated directed graph $D(A)$ has a path connecting vertex i to vertex j. It follows that A is irreducible if and only if for each i and j there exists a path in $D(A)$ connecting i to j; in other words, if and only if $D(A)$ is strongly connected. ■

Graphical methods can be used for determining whether a given nonnegative matrix is primitive or not, and for finding the index of imprimitivity of an irreducible matrix.

Definition 3.4. Let D be a strongly connected directed graph. The g.c.d. of the lengths of all cycles in D is called the *index of imprimitivity of D*.

Lemma 3.1. *Let D be a strongly connected graph with index of imprimitivity k, and let k_i be the g.c.d. of the lengths of all cycles of D through vertex i. Then $k_i = k$.*

Proof. Clearly, $k|k_i$. We show that $k_i|k$. Let C_j be any cycle in D. Let its length be m_j, and suppose that it passes through vertex j. Since D is strongly connected it contains a path P_{ij} connecting vertex i to vertex j, and a path P_{ji} connecting vertex j to vertex i. Let the lengths of these paths be s_{ij} and s_{ji}, respectively. Now, the path consisting of P_{ij} and P_{ji} and the path consisting of P_{ij}, C_j, and P_{ji} are cycles through vertex i. The lengths of these cycles are $s_{ij} + s_{ji}$ and $s_{ij} + m_j + s_{ji}$. Since k_i divides both $s_{ij} + s_{ji}$ and $s_{ij} + m_j + s_{ji}$, it must divide m_j. In other words, k_i divides the length of every cycle in D. Thus k_i divides k, and we can conclude that $k_i = k$. ■

Theorem 3.3. *The index of imprimitivity of an irreducible matrix is equal to the index of imprimitivity of the associated directed graph.*

Proof. Let h be the index of imprimitivity of an irreducible $n \times n$ matrix $A = (a_{ij})$, and let k be the index of imprimitivity of the associated strongly connected, directed graph $D(A)$. Consider the cycles through vertex i. Let M_i be the set of lengths of these cycles. Then, by Lemma 3.1,

$$k = \text{g.c.d.}\{m_t | m_t \in M_i\}. \tag{1}$$

We show that M_i is closed under addition. Let m_1 and m_2 be any integers in M_i. Then $a_{ii}^{(m_1)} > 0$ and $a_{ii}^{(m_2)} > 0$, by Corollary 3.1. It follows that

$$a_{ii}^{(m_1+m_2)} = \sum_{t=1}^{n} a_{it}^{(m_1)} a_{ti}^{(m_2)} \geq a_{ii}^{(m_1)} a_{ii}^{(m_2)} > 0.$$

Hence, by Corollary 3.1, there exists a cycle of length $m_1 + m_2$ through vertex i, and therefore $m_1 + m_2 \in M_i$. Thus M_i is closed under addition, and, by a well-known theorem of Schur, it must contain all but a finite number of multiples of k. Therefore $a_{ii}^{(kt)} > 0$ for all sufficiently large integers t. On the other hand, if s is not a multiple of k, then it follows from (1), the definition of M_i, and Corollary 3.1, that $a_{ii}^{(s)} = 0$. Since i was any vertex of $D(A)$ we can conclude that $a_{ii}^{(s)} > 0$ for all sufficiently large s and $i = 1, 2, \ldots, n$, if and only if s is a multiple of k.

Matrix A is irreducible with index of imprimitivity h. Hence if $h > 1$, then, by Theorem 3.1, Chapter III, there exists a permutation matrix P such that $P^{\mathrm{T}}AP$ is in the Frobenius form with h nonzero superdiagonal blocks. By Lemma 4.1 and Corollary 4.1, Chapter III, $(P^{\mathrm{T}}AP)^h$ is a direct sum of primitive matrices. Hence, for sufficiently large t, $(P^{\mathrm{T}}AP)^{ht} = P^{\mathrm{T}}A^{ht}P$ is a direct sum of positive matrices. It follows that for all sufficiently large t, $a_{ii}^{(ht)} > 0$, $i = 1, 2, \ldots, n$. On the other hand, if s is not a multiple of h, then all the main diagonal entries in $(P^{\mathrm{T}}AP)^s$, that is, of A^s, are zero. If $h = 1$, then A is primitive and, for all sufficiently large t, the matrix $A^t = A^{ht}$ is positive. We can conclude that in either case, for sufficiently large s, $a_{ii}^{(s)} > 0$ if and only if s is a multiple of h. Now it follows that $h = k$. ■

Since a nonnegative matrix has a nonzero main diagonal entry if and only if the associated directed graph has a loop, that is, a cycle of length 1, we have the following result.

Corollary 3.2. *An irreducible matrix with a nonzero main diagonal is primitive* (cf. Corollary 1.1, Chapter III).

Theorem 3.3 provides a useful method for finding the index of imprimitivity of an irreducible matrix. It is doubtful, however, whether this method is more efficient than the one provided by Theorem 3.1, Chapter III. The task here is somewhat facilitated by the observation (see Problem 12) that the index of imprimitivity of a directed graph is equal to the g.c.d. of the lengths of all the circuits in the graph.

Example 3.3. Use graph methods to show that

$$A = \begin{bmatrix} 0 & 1 & 0 & 0 & 0 & 0 \\ 1 & 0 & 0 & 0 & 1 & 0 \\ 0 & 1 & 0 & 0 & 0 & 0 \\ 1 & 0 & 1 & 0 & 0 & 0 \\ 0 & 0 & 0 & 1 & 0 & 1 \\ 0 & 0 & 1 & 0 & 1 & 0 \end{bmatrix}$$

is irreducible. Find its index of imprimitivity.

We construct the associated directed graph of A.

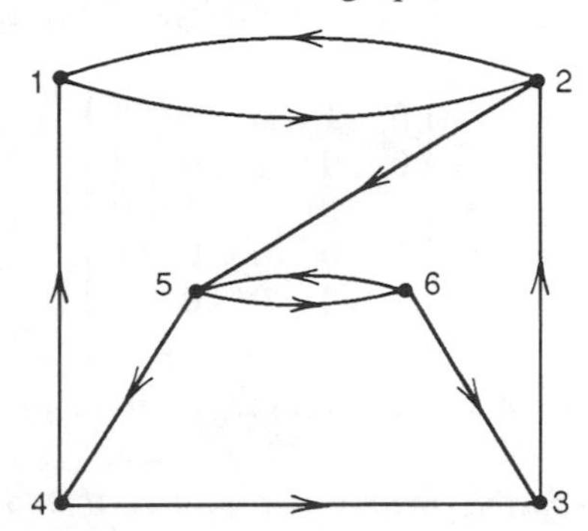

We construct the associated directed graph of A.

We check that there is a path connecting every pair of vertices, and thus that the graph is strongly connected. Clearly, it is sufficient to check that vertex 1, say, is connected to every other vertex and each of these vertices is connected to vertex 1. It then follows that A is irreducible. We observe that there is a cycle of length 2 through vertex 1, and that there is no circuit of odd length. We can conclude that the index of imprimitivity of the graph (and thus the index of imprimitivity of A) is 2. It is clear from the graph that if, e.g., the arc (2, 6) were added to the graph, the graph would have a cycle of length 3, its index of imprimitivity would be 1, and its adjacency matrix would be primitive. [Perhaps it is not immediately obvious by inspection that replacing the zero in the (2, 6) position by 1 would render the resulting matrix primitive.] ■

We conclude this section with a characterization of primitive matrices due to Lewin [18].

Theorem 3.4. *If $A = (a_{ij})$ is an irreducible matrix and*

$$a_{ij}a_{ij}^{(2)} > 0$$

for some (i, j), then A is primitive.

Proof. By Corollary 3.1, in the associated directed graph $D(A)$ vertex i is connected to vertex j by paths of lengths 1 and 2. Since A is irreducible, and therefore, by Theorem 3.2, $D(A)$ is strongly connected, there exists a path connecting vertex j to vertex i of length s, say. Then $D(A)$ contains cycles of lengths $s + 1$ and $s + 2$. Since g.c.d.$(s + 1, s + 2) = 1$ it follows, by Theorem 3.3, that A is primitive. ■

Theorem 3.4 can be expressed in the following form: *If A is an irreducible matrix and the Hadamard product $A * A^2$ is nonzero, then A is primitive.* [Recall that the Hadamard product of two $m \times n$ matrices $X = (x_{ij})$ and $Y = (y_{ij})$ is the $m \times n$ matrix whose (i, j) entry is $x_{ij}y_{ij}$.]

Note that the converse of Theorem 3.4 is false. For example, the matrix

$$A = \begin{bmatrix} 0 & 1 & 0 & 1 & 0 \\ 0 & 0 & 1 & 0 & 0 \\ 0 & 0 & 0 & 1 & 0 \\ 0 & 0 & 0 & 0 & 1 \\ 1 & 0 & 0 & 0 & 0 \end{bmatrix}$$

is primitive, although $A * A^2 = 0$. However, we have the following result.

Theorem 3.5. *An irreducible matrix is primitive if and only if there exists a positive integer q such that $A^q * A^{q+1}$ is nonzero.*

Proof. The sufficiency is proved by the method used in the proof of Theorem 3.4. The necessity is an obvious consequence of Theorem 2.1, Chapter III. ■

4.4. FULLY INDECOMPOSABLE MATRICES

In the preceding chapters we saw that the spectral properties of a nonnegative matrix can be made more apparent by a suitable permutation of its rows and the same permutation of its columns. In studying combinatorial properties of a nonnegative matrix we can usually go a step further: We can permute its rows and columns independently without affecting its essential combinatorial characteristics. For example, permuting rows of an incidence matrix corresponds to relabelling the subsets in the configuration whereas a permutation of columns is equivalent to relabelling its elements. We shall call two matrices A and B *permutation equivalent*, or *p-equivalent*, if there exist permutation matrices P and Q such that $A = PBQ$. In the spectral theory of nonnegative matrices the key concept is that of an irreducible matrix, that is, a matrix which is not cogredient to subdirect sum of smaller matrices. In combinatorial theory the equivalent concept is that of a fully indecomposable matrix, that is, a matrix which is not p-equivalent to any subdirect sum.

Definition 4.1. An $n \times n$ nonnegative matrix is said to be *partly decomposable* if it contains an $s \times (n - s)$ zero submatrix. In other words, a matrix is partly decomposable if it is p-equivalent to a matrix of the form

$$\begin{bmatrix} X & Y \\ 0 & Z \end{bmatrix},$$

where X and Z are square. If an $n \times n$ nonnegative matrix contains no $s \times (n - s)$ zero submatrix, it is said to be *fully indecomposable*. Thus a nonnegative matrix is fully indecomposable if it is not partly decomposable. The 1×1 zero matrix is, by definition, partly decomposable, whereas a nonzero 1×1 matrix is fully indecomposable.

An important characterization of fully indecomposable matrices in terms of permanents is contained in the following theorem.

Theorem 4.1 (Marcus and Minc [20]). *A nonnegative $n \times n$ matrix A, $n \geq 2$, is fully indecomposable if and only if*

$$\operatorname{per}(A(i|j)) > 0,$$

for all i and j.

Proof. By the Frobenius–König theorem (Theorem 2.1), $\operatorname{per}(A(h|k)) = 0$ for some h and k if and only if the submatrix $A(h|k)$, and thus the matrix A, contains an $s \times t$ zero submatrix with $s + t = (n-1) + 1 = n$. Hence $\operatorname{per}(A(h|k)) = 0$ for some h and k if and only if A is partly decomposable. ∎

Corollary 4.1. *Every positive entry of a fully indecomposable matrix lies on a positive diagonal.*

Recall that E_{ij} denotes the $n \times n$ matrix with 1 in the (i, j) position and 0's elsewhere.

Corollary 4.2. *If A is a fully indecomposable nonnegative matrix, and c is a nonzero real number, then for every i and j,*

$$\operatorname{per}(A + cE_{ij}) > \operatorname{per}(A)$$

or

$$\operatorname{per}(A + cE_{ij}) < \operatorname{per}(A),$$

according as $c > 0$ or $c < 0$.

For, $\operatorname{per}(A + cE_{ij}) = \operatorname{per}(A) + c\operatorname{per}(A(i|j))$ and, by Theorem 4.1, $\operatorname{per}(A(i|j)) > 0$.

A stronger result is possible in the case of a fully indecomposable (0,1)-matrix.

Corollary 4.3. *If A is a fully indecomposable (0,1)-matrix, then*

$$\operatorname{per}\left(A + \sum_{t=1}^{m} E_{i_t j_t}\right) \geq \operatorname{per}(A) + m.$$

Proof. Since A is a (0,1)-matrix, Theorem 4.1 implies that

$$\operatorname{per}(A(i|j)) \geq 1,$$

for all i and j. Therefore

$$\begin{aligned}\operatorname{per}(A + E_{i_1 j_1}) &= \operatorname{per}(A) + \operatorname{per}(A(i_1|j_1)) \\ &\geq \operatorname{per}(A) + 1.\end{aligned}$$

Clearly, $A + E_{i_1 j_1}$ is fully indecomposable. The result now follows by induction on m. ■

Theorem 4.2. *Let*

$$A = \begin{bmatrix} A_1 & B_1 & 0 & \cdots & & 0 \\ 0 & A_2 & B_2 & & & \vdots \\ \vdots & & & \ddots & & 0 \\ 0 & 0 & & & A_{r-1} & B_{r-1} \\ B_r & 0 & & \cdots & 0 & A_r \end{bmatrix} \tag{1}$$

be a nonnegative $n \times n$ matrix, where A_i is a fully indecomposable $n_i \times n_i$ matrix and $B_i \neq 0$, $i = 1, \ldots, r$. Then A is fully indecomposable.

Proof (Edgar [7]). Suppose that A is partly decomposable, that is, $A[\alpha|\beta] = 0$ for some $\alpha \in Q_{s,n}$ and $\beta \in Q_{t,n}$, where $s + t = n$. Let s_j of rows α and t_j of columns β intersect the submatrix A_j, $j = 1, \ldots, r$. Then $s_1 + s_2 + \cdots + s_r = s \geq 1$, so that at least one of the s_j must be positive. Similarly, at least one of the t_j is not zero. Now, since each A_j is fully indecomposable and contains an $s_j \times t_j$ zero submatrix (unless either $s_j = 0$ or $t_j = 0$), we must have $s_j + t_j \leq n_j$, where equality can hold only if $s_j = 0$ or $t_j = 0$. But

$$\begin{aligned} n &= s + t \\ &= \sum_{j=1}^{r} s_j + \sum_{j=1}^{r} t_j \\ &= \sum_{j=1}^{r} (s_j + t_j) \\ &\leq \sum_{j=1}^{r} n_j \\ &= n, \end{aligned}$$

and thus $s_j + t_j = n_j$ for every j. It follows that either $s_j = 0$ or $t_j = 0$ for

$j = 1, \ldots, r$. But not all the s_j nor all the t_j can be zero. Therefore there must exist an integer k such that $s_k = n_k$ and $t_{k+1} = n_{k+1}$ (subscripts reduced modulo r). But then B_k is a submatrix of a zero submatrix, contradicting our hypotheses. ■

Definition 4.2. A nonnegative matrix is called *doubly stochastic* if all its row and column sums are 1.

Clearly, a doubly stochastic matrix must be square. We shall study doubly stochastic matrices in detail in the next chapter.

Definition 4.3. A nonnegative matrix is said to have a *doubly stochastic pattern* if it has the same zero pattern as a doubly stochastic matrix. ■

For example, the matrix

$$\begin{bmatrix} 1 & 1 & 1 \\ 1 & 1 & 0 \\ 1 & 0 & 1 \end{bmatrix}$$

has a doubly stochastic pattern, since it has the same zero pattern as the doubly stochastic matrix

$$\begin{bmatrix} \frac{1}{2} & \frac{1}{4} & \frac{1}{4} \\ \frac{1}{4} & \frac{3}{4} & 0 \\ \frac{1}{4} & 0 & \frac{3}{4} \end{bmatrix}.$$

On the other hand, the matrix

$$\begin{bmatrix} 1 & 1 & 1 \\ 1 & 1 & 0 \\ 0 & 0 & 1 \end{bmatrix} \tag{2}$$

does not have a doubly stochastic pattern. For, if any doubly stochastic matrix had the same zero pattern as (2), its only nonzero entry in the third row would have to be 1. But then its (1, 3) entry could not be positive.

Theorem 4.3. *A fully indecomposable matrix has a doubly stochastic pattern.*

Proof. Let $A = (a_{ij})$ be an $n \times n$ fully indecomposable matrix. Then, by Theorem 4.1, $\operatorname{per}(A(i|j)) > 0$ for all i, j. Let $S = (s_{ij})$ be the $n \times n$ matrix defined by

$$s_{ij} = a_{ij}\operatorname{per}(A(i|j))/\operatorname{per}(A), \qquad i, j = 1, \ldots, n.$$

Clearly, s is nonnegative and it has the same zero pattern as A. Also for $i = 1, \ldots, n$,

$$\begin{aligned}\sum_{j=1}^{n} s_j &= \frac{1}{\operatorname{per}(A)} \sum_{j=1}^{n} a_{ij} \operatorname{per}(A(i|j)) \\ &= \frac{1}{\operatorname{per}(A)} \operatorname{per}(A) \\ &= 1,\end{aligned}$$

and similarly for $j = 1, \ldots, n$,

$$\begin{aligned}\sum_{i=1}^{n} s_{ij} &= \frac{1}{\operatorname{per}(A)} \sum_{i=1}^{n} a_{ij} \operatorname{per}(A(i|j)) \\ &= 1.\end{aligned}$$

Hence S is doubly stochastic and thus A has a doubly stochastic pattern. ■

Note that the converse of Theorem 4.3 is not true (Problem 16).

The following simple result [4] illustrates the connection between the concepts of full indecomposability and irreducibility.

Theorem 4.4. *A nonnegative matrix is fully indecomposable if and only if it is p-equivalent to an irreducible matrix with a positive main diagonal.*

Proof. Let A be an irreducible $n \times n$ matrix with a positive main diagonal. If A were partly decomposable, then it would contain a zero submatrix $A[i_1, i_2, \ldots, i_s | j_{s+1}, j_{s+2}, \ldots, j_n]$, $i_1 < i_2 < \cdots < i_s$, $j_{s+1} < j_{s+2} < \cdots < j_n$, where $\{i_1, i_2, \ldots, i_s\}$ and $\{j_{s+1}, j_{s+2}, \ldots, j_n\}$ would have to be disjoint since A has no zero entries on its main diagonal. But this would contradict the fact that A is irreducible (see Problem 6, Chapter III).

Conversely, a fully indecomposable matrix contains, by Corollary 4.1, a positive diagonal and is therefore p-equivalent to a fully indecomposable, and thus irreducible, matrix with a positive main diagonal. ■

4.5. NEARLY DECOMPOSABLE AND NEARLY REDUCIBLE MATRICES

Corollary 4.2 implies that if a positive entry in a fully indecomposable $(0,1)$-matrix is replaced by a zero, then, provided that the matrix remains fully indecomposable, its permanent decreases by at least 1, that is, the permanent of the original matrix exceeds the permanent of the resulting matrix by at least 1. If the latter were known we would have a lower bound for the permanent of the original matrix. If it is not known or it is difficult to estimate, we may continue the process of replacing 1's by 0's until we arrive at a more tractable fully indecomposable matrix, and then apply Corollary 4.3 in order to obtain a lower bound for the permanent of the original matrix. We shall now consider the class of matrices obtained by carrying out such a "stripping" process on fully indecomposable matrices as far as possible.

Definition 5.1. (a) A fully indecomposable nonnegative matrix $A = (a_{ij})$ is said to be *nearly decomposable* if for each $a_{hk} > 0$, the matrix $A - a_{hk}E_{hk}$ is partly decomposable.

(b) An irreducible nonnegative matrix $A = (a_{ij})$ is called *nearly reducible* if for each positive a_{hk}, the matrix $A - a_{hk}E_{hk}$ is reducible. It is convenient here to regard a 1×1 zero matrix as nearly reducible (and therefore irreducible).

(c) A strongly connected directed graph is called *minimally connected* if it ceases to be strongly connected when any of its arcs is deleted.

(d) A strongly connected directed graph with no loops is called a *rosette* if it has at most one vertex to which more than two arcs are incident.

(e) If W is a subset of vertices of a directed graph D, then the *shrink* of W is obtained from D by deleting the arcs joining any two vertices of W and by identifying all the vertices of W with a single one of them.

Note that the shrink of W may not be a directed graph since it may have multiple arcs joining the same (ordered) pair of vertices. However, if no pair of vertices is joined by more than one arc, then the shrink is a directed graph.

Lemma 5.1. (a) *A directed graph with at least two vertices is minimally connected if and only if its adjacency matrix is nearly reducible.*

(b) *A rosette (and thus a circuit) is a minimally connected graph.*

(c) *The adjacency matrix of a circuit is a full-cycle permutation matrix.*

The above propositions are quite obvious. We leave their proofs to the reader (Problem 21).

Lemma 5.2. *Let A be a matrix in form* (1), *Section* 4.4, *where the A_i are fully indecomposable and the B_i are nonzero. If A is nearly decomposable, then each of the A_i is nearly decomposable.*

Proof. If A_t were not nearly decomposable, a positive entry in A_t could be replaced by a zero, and the resulting block would still be fully indecomposable. But then, by Theorem 4.2, matrix A with the same positive entry replaced by a zero would still be fully indecomposable, contradicting the fact that A is nearly decomposable. ■

Lemma 5.3 (Berge [2]). *If W is the set of vertices of a strongly connected subgraph of a minimally connected directed graph D, then the shrink of W is also a minimally connected directed graph.*

Proof. We first show that the shrink of W is a directed graph, that is, it cannot contain multiple arcs. For, if it were not the case, there would exist in D pairs of arcs (i, j) and (i, k) [or (j, i) and (k, i)] with $i \notin W$ and $j, k \in W$. But then the strongly connected graph D would not be minimally connected, since the directed graph obtained from D by deleting one of the two arcs would again be strongly connected.

Next we assert that the shrink of W is minimally connected. Clearly, it is strongly connected. Also, the deletion of any of its arcs cannot result in a strongly connected graph. Otherwise, the deletion of the same arc in D would also result in a strongly connected graph, contradicting the fact that D is minimally connected. ∎

The following theorem gives a remarkably simple canonical form for nearly reducible and nearly decomposable matrices. The result is due to Hartfiel [13]. His canonical form for nearly decomposable matrices is a substantial simplification of the canonical form obtained by Sinkhorn and Knopp [28].

Theorem 5.1. *Let A be an $n \times n$ nearly decomposable (nearly reducible) matrix, $n > 1$. Then A is p-equivalent (cogredient) to a matrix of the form*

$$\begin{bmatrix} A_1 & E_1 & 0 & \cdots & 0 & 0 \\ 0 & A_2 & E_2 & \cdots & 0 & 0 \\ \vdots & & & \ddots & & \vdots \\ 0 & 0 & 0 & & A_{s-1} & E_{s-1} \\ E_s & 0 & 0 & \cdots & 0 & A_s \end{bmatrix}, \tag{1}$$

where $s \geq 2$, each E_i has exactly one positive entry, each A_i is nearly decomposable (nearly reducible), and all the A_i, except possibly A_s, are 1×1.

Proof. We first prove the theorem for nearly reducible matrices. Let D be the directed graph associated with A. By Lemma 5.1(a), graph D is minimally connected. If D is a Hamiltonian circuit, then by Lemma 5.1(c) the adjacency matrix of D is a full-cycle permutation matrix, and A is cogredient to a matrix of the form (1). If D is not a Hamiltonian circuit, we shrink any of its circuits. By Lemma 5.3, the resulting graph D_1 is minimally connected. Let v_1 be the vertex of D_1 which replaced the vertices of the shrunk circuit. If D_1 is not a Hamiltonian circuit, we shrink any circuit through v_1 and obtain a minimally connected graph D_2; let v_2 be the vertex which replaced the vertices of the shrunk circuit through v_1. We continue this shrinking process until we arrive at a minimally connected graph D_m which is a Hamiltonian circuit of length s. Let v_m be the vertex to which all the circuits in the preceding stages shrank. The vertices of D which shrank to v_m form a subgraph of D which is a rosette. The remaining $s - 1$ vertices were not involved in the shrinking process.

Relabel them as $1, 2, \ldots, s-1$, label vertex v_m as s, and relabel the remaining vertices of D as $s+1, s+2, \ldots, n$, in any order. Then C, the adjacency matrix of D with its vertices relabelled as specified above, is of the form (1), where $A_1 = A_2 = \cdots = A_{s-1} = 0$, $E_1 = E_2 = \cdots = E_{s-2} = 1$, E_{s-1} is $1 \times (n-s+1)$ with 1 as its first entry and 0's elsewhere, and E_s is $(n-s+1) \times 1$ with 1 as its top entry and 0's elsewhere. Matrix A is cogredient to a matrix with the same zero pattern as C. This completes the proof for the case when A is nearly reducible.

Now, suppose that A is nearly decomposable. By Theorem 4.4, matrix A is p-equivalent to an irreducible matrix $\bar{A}$ with a positive main diagonal. Since the irreducibility of a matrix does not depend on its main diagonal entries, $\bar{A}$ is a sum of an irreducible matrix B with a zero diagonal and a nonsingular nonnegative diagonal matrix L. Clearly, B is nearly reducible. Hence, by what we have already proved, there exists a permutation matrix P such that $P^{\mathrm{T}}BP$ is of the form (1), where the main diagonal blocks are nearly reducible and each E_i has exactly one positive entry. Again, by Theorem 4.4, $P^{\mathrm{T}}BP + P^{\mathrm{T}}LP = P^{\mathrm{T}}\bar{A}P$ is of the form (1), where the main diagonal blocks are now fully indecomposable, and each E_i has exactly one positive entry. Since $P^{\mathrm{T}}\bar{A}P$ is nearly decomposable, then, by Lemma 5.2, each main diagonal block of $P^{\mathrm{T}}\bar{A}P$ is also nearly decomposable. Matrix $P^{\mathrm{T}}\bar{A}P$ is p-equivalent to A. ∎

We use Theorem 5.1 to evaluate the maximum number of positive entries in a nearly reducible or a nearly decomposable matrix. We start with a lemma which may be of interest in itself.

Lemma 5.4. *If $A = (a_{ij})$ is a nearly decomposable matrix in canonical form* (1), *then A_s cannot be 2×2.*

Proof. Suppose that A_s is 2×2. We can assume without loss of generality that A is a (0,1)-matrix, and that $E_{n-2} = [1 \quad 0]$ whereas $E_{n-1} = [0 \quad 1]^{\mathrm{T}}$. Since A_s is a fully indecomposable 2×2 matrix it must be positive. But then A cannot be nearly decomposable since $A - E_{n,n-1} = I_n + P_n$ is fully indecomposable by Theorem 4.2. (Here P_n denotes the full-cycle permutation matrix with 1's in the superdiagonal.) ∎

Lemma 5.5. (a) *Let B_n be the $n \times n$ (0,1)-matrix, $n \geq 2$,*

$$B_n = \sum_{j=2}^{n} E_{1j} + \sum_{i=2}^{n} E_{i1} = \begin{bmatrix} 0 & 1 & 1 & \cdots & 1 \\ 1 & & & & \\ 1 & & 0 & & \\ \vdots & & & & \\ 1 & & & & \end{bmatrix}.$$

Then B_n is nearly reducible.

(b) *Let C_n be the $n \times n$ (0,1)-matrix, $n \geq 3$,*

$$C_n = B_n + \sum_{i=2}^{n} E_{ii} = \begin{bmatrix} 0 & 1 & 1 & \cdots & 1 \\ 1 & 1 & 0 & & 0 \\ 1 & 0 & 1 & & 0 \\ \vdots & & & \ddots & \vdots \\ 1 & 0 & 0 & \cdots & 1 \end{bmatrix}.$$

Then C_n is nearly decomposable, and

$$\operatorname{per}(C_n) = -\det(C_n) = n - 1.$$

(Note that the matrix C_n occurs in a celebrated theorem of de Bruijn and Erdös [5].)

Proof. (a) Clearly, B_n is nearly reducible. In fact, its associated directed graph is a rosette.

(b) Matrix $B_n + I_n = C_n + E_{11}$ is fully indecomposable, by Theorem 4.4, and so is C_n since the additional zero in the (1,1) position cannot be an entry in any $s \times (n - s)$ zero submatrix for $n \geq 2$. Also, C_n is nearly decomposable: If any of its positive entries is replaced by a zero, then the resulting matrix contains either a $1 \times (n - 1)$ or an $(n - 1) \times 1$ zero submatrix.

Subtract the sum of the last $n - 1$ columns of C_n from its first column. The resulting matrix is triangular and its main diagonal product is $-(n - 1)$ which is equal to $\det(C_n)$.

To compute the permanent of C_n use induction on n. The permanent of C_3 clearly is 2. Assume that $n > 3$ and that $\operatorname{per}(C_{n-1}) = n - 2$. Then

$$\begin{aligned} \operatorname{per}(C_n) &= \operatorname{per}(C_n(1|n)) + \operatorname{per}(C_n(n|n)) \\ &= 1 + n - 2 \\ &= n - 1. \quad \blacksquare \end{aligned}$$

Let $\sigma(X)$ denote the sum of all entries in matrix X.

Theorem 5.2 (Minc [23]). *If A is a nearly decomposable $n \times n$ (0,1)-matrix, $n \geq 3$, then*

$$\sigma(A) \leq 3(n - 1). \tag{2}$$

Equality holds in (2) if and only if A is p-equivalent to C_n.

Proof. Let $\tilde{A}$ be a matrix in the form (1) p-equivalent to A. Then

$$\begin{aligned}\sigma(A)=\sigma(\tilde{A}) &= \sum_{i=1}^{s}\sigma(E_i)+\sum_{i=1}^{s}\sigma(A_i)\\ &= s+(s-1)+\sigma(A_s)\\ &= 2s-1+\sigma(A_s).\end{aligned}$$

Use induction on n. If A_s is 1×1, then $s=n$, $\tilde{A}=I_n+P_n$, and

$$\begin{aligned}\sigma(A) &= 2n-1+1\\ &= 2n\\ &\le 3n-3,\end{aligned}$$

where equality holds if and only if $n=3$. It is easily seen that the matrix I_3+P_3 is p-equivalent to C_3.

By Lemma 5.4, submatrix A_s cannot be 2×2. Assume that $n-s+1\ge 3$ and use the induction hypothesis:

$$\begin{aligned}\sigma(A) &= 2s-1+\sigma(A_s)\\ &\le 2s-1+3((n-s+1)-1) &\quad (3)\\ &= 3n-s-1\\ &\le 3(n-1), &\quad (4)\end{aligned}$$

since $s\ge 2$. Equality holds in (2) if and only if inequalities (4) and (3) are equalities, that is, if and only if $s=2$ and $\sigma(A_s)=3((n-s+1)-1)$, which implies by the induction hypothesis that submatrix A_s (that is A_2) is p-equivalent to C_{n-1}. It follows that A is p-equivalent to

$$C=\left[\begin{array}{c|cccc} 1 & 1 & 0 & \cdots & 0\\ \hline 1 & & & & \\ 0 & & & & \\ \vdots & & & C_{n-1} & \\ 0 & & & & \end{array}\right], \quad (5)$$

which obviously is p-equivalent to C_n. Note that the 1's in E_1 and E_2 must be in positions $(1,2)$ and $(2,1)$ in C. For, if E_1 had 1 in position $(1,j)$, $j\ge 3$, then $C-E_{2j}$ would be fully indecomposable contradicting the fact that C is nearly decomposable. For a similar reason E_2 cannot have its 1 in position $(i,1)$, $i\ge 3$. ■

Corollary 5.1. *A nearly decomposable nonnegative $n\times n$ matrix must have at least n^2-3n+3 zero entries.*

Example 5.1. Show that for any $n \geq 3$ and any N, $2n \leq N \leq 3(n-1)$, there exists a nearly decomposable $n \times n$ (0, 1)-matrix with N positive entries.

Let G_N^n be the required matrix. If $N = 3(n-1)$ take $G_N^n = C_n$, the matrix in Lemma 5.5(b). If $N = 2n$ choose $G_N^n = I_n + P_n$. If $2n < N < 3(n-1)$ let

$$G_N^n = \left[\begin{array}{cccccc|cccc} 1 & 1 & 0 & \cdots & & 0 & 0 & & & \\ 0 & 1 & 1 & \cdots & & 0 & & & & \\ \vdots & & & \ddots & & \vdots & \vdots & & 0 & \\ & & & & 1 & 1 & 0 & & & \\ 0 & & & \cdots & 0 & 1 & 1 & 0 & \cdots & 0 \\ \hline 1 & 0 & & \cdots & & 0 & & & & \\ 0 & & & & & & & & & \\ \vdots & & & 0 & & & & C_{n-s+1} & & \\ 0 & & & & & & & & & \end{array}\right],$$

where $s = 3n - N - 1$ and the leading block is $(s-1) \times (s-1)$. Note that $3 \leq n - s + 1 \leq n - 2$. Clearly, G_N^n is nearly decomposable, and

$$\begin{aligned} \sigma(G_N^n) &= 2s - 1 + \sigma(C_{n-s+1}) \\ &= 2s - 1 + 3(n-s) \\ &= 3n - s - 1 \\ &= 3n - 1 - (3n - 1 - N) \\ &= N. \quad \blacksquare \end{aligned}$$

Theorem 5.3. *If A is a nearly reducible $n \times n$ (0, 1)-matrix, $n \geq 2$, then*

$$\sigma(A) \leq 2(n-1).$$

The proof of Theorem 5.3 is similar to that of the preceding theorem; it is left as an exercise for the reader (Problem 23).

4.6. BOUNDS FOR PERMANENTS OF (0, 1)-MATRICES

In Section 4.1 we introduced the concepts of the incidence matrix of a configuration of subsets, and of a system of distinct representatives (SDR) of the configuration. We saw that an SDR corresponds to a positive diagonal in the incidence matrix, and therefore the number of SDRs in a configuration is equal to the permanent of the incidence matrix of the configuration. Unfortunately, there is no efficient algorithm for computing permanents. In general, it is impossible to compute the permanent of a large matrix, even with the use of computers. In these cases we have to be satisfied with bounds for permanents.

We first restate the Frobenius-König theorem (Theorem 2.1) as a condition for the existence of an SDR.

Theorem 6.1. *A configuration of m subsets of an n-set, $m \leq n$, contains an SDR if and only if the incidence matrix of the configuration does not contain an $s \times (n - s + 1)$ zero submatrix, $1 \leq s \leq m$.*

Corollary 6.1 (Minc [24]). *If in a configuration of m subsets of an n-set, $m \leq n$, every subset contains at least m elements, then the configuration has an SDR. In other words, if each row sum of an $m \times n$ $(0,1)$-matrix, $m \leq n$, is greater than or equal to m, then its permanent is greater than or equal to* 1.

The corollary is an obvious consequence of the Frobenius–König theorem. For, if the permanent of the matrix did vanish, then the matrix would contain an $s \times (n - s + 1)$ zero submatrix, which is impossible since the matrix has at most $n - m$ zero columns and $s \leq m$.

The first significant lower bound for the number of SDRs in a configuration was obtained in 1948 by Hall [11]. We state it here, in a somewhat extended form due to Mann and Ryser [19], in terms of permanents of (0, 1)-matrices.

Theorem 6.2. *Let A be an $m \times n$ $(0,1)$-matrix, $m \leq n$, with at least t 1's in each row. If $t \geq m$, then*

$$\operatorname{Per}(A) \geq t!/(t-m)!.$$

If $t \leq m$ and $\operatorname{Per}(A) > 0$, *then*

$$\operatorname{Per}(A) \geq t!.$$

Proof. By virtue of Corollary 6.1, we can assume that $\operatorname{Per}(A) > 0$ for all values of t, $0 < t \leq n$. Use induction on m. If $m = 1$, then $t \geq m$ and $\operatorname{Per}(A) = t = t!/(t-m)!$. Now, assume that $m > 1$ and that the theorem holds for all matrices with fewer than m rows. Since the permanent of A is positive, the matrix cannot contain a $k \times (n - k + 1)$ zero matrix and thus every $k \times n$ submatrix of A contains at least k nonzero columns. Suppose first that for some h, $1 \leq h \leq m - 1$, A contains an $h \times n$ submatrix with exactly $n - h$ zero columns, that is, A is p-equivalent to

$$h\{\left[\begin{array}{c|c} \overbrace{B}^{h} & 0 \\ \hline C & D \end{array}\right]. \tag{1}$$

In each of the first h rows of the matrix the t positive entries must be contained in B. Thus each row of B has at least t 1's and $t \leq h \leq m - 1$.

Also,

$$\text{Per}(A) = \text{Per}(B) \cdot \text{Per}(D) > 0.$$

Hence $\text{Per}(B) > 0$ and $\text{Per}(D) > 0$. By the induction hypothesis, $\text{Per}(B) \geq t!$ and therefore

$$\begin{aligned}\text{Per}(A) &= \text{Per}(B) \cdot \text{Per}(D) \\ &\geq t!\text{Per}(D) \\ &\geq t!.\end{aligned}$$

If A is not p-equivalent to a matrix in the form (1), then every $k \times n$ submatrix of A, $1 \leq k \leq m-1$, must contain at least $k+1$ nonzero columns. Thus every $k \times (n-1)$ submatrix of A has at least k nonzero columns, and, by Theorem 2.1, every $(m-1) \times (n-1)$ submatrix $A(s|t)$ has a positive permanent. Also, every row of $A(s|t)$ has at least $t-1$ ones. Hence, by the induction hypothesis,

$$\text{Per}(A(s|t)) \geq \begin{cases} (t-1)!, & \text{if } t-1 \leq m-1, \\ (t-1)!/(t-m)!, & \text{if } t-1 \geq m-1. \end{cases} \tag{2}$$

But $t-1 \leq m-1$ if $t \leq m$, and $t-1 \geq m-1$ if $t \geq m$. Hence if $t \leq m$, then

$$\begin{aligned}\text{Per}(A) &= \sum_{j=1}^{n} a_{1j}\text{Per}(A(1|j)) \\ &\geq \sum_{j=1}^{n} a_{1j}(t-1)! \\ &= (t-1)! \sum_{j=1}^{n} a_{1j} \\ &\geq t!,\end{aligned}$$

since $\sum_{j=1}^{n} a_{1j} \geq t$. Similarly, if $t \geq m$, then

$$\begin{aligned}\text{Per}(A) &\geq \sum_{j=1}^{n} a_{1j}\frac{(t-1)!}{(t-m)!} \\ &= \frac{(t-1)!}{(t-m)!}\sum_{j=1}^{n} a_{1j} \\ &\geq \frac{t!}{(t-m)!} \quad \text{(see also Problem 32).} \quad \blacksquare\end{aligned}$$

Note that the condition $\text{Per}(A) > 0$ in the statement of Theorem 6.2 is essential. Even if every row sum of an $m \times n$ $(0,1)$-matrix is $n-1$, its permanent may vanish (Problem 33). Of course, it may not be easy to decide whether the permanent of a (0,1)-matrix is zero. In fact, it may take as many as $O(n^{5/2})$ computational steps [14]. If the matrix happens to be square and fully indecomposable, then its permanent is positive, but the determination of whether a matrix is fully indecomposable or not, is, in general, of the same order of difficulty as that of whether its permanent is zero. However, for permanents of fully indecomposable (0,1)-matrices other lower bounds are available.

Theorem 6.3 (Minc [22]). *If $A = (a_{ij})$ is a fully indecomposable $n \times n$ (0,1)-matrix, then*

$$\operatorname{per}(A) \geq \sigma(A) - 2n + 2, \tag{3}$$

where $\sigma(A)$ denotes the sum of all entries in A.

Proof. First, suppose that A is nearly decomposable, and use induction on n. For nearly decomposable matrices with $n = 1, 2, 3$, inequality (3) becomes equality (Problem 34). Assume the theorem holds for nearly decomposable $m \times m$ (0,1)-matrices with $3 < m < n$. Let P and Q be permutation matrices such that $B = PAQ$ is in the canonical form (1), Section 4.5. Suppose that A_s is $n_s \times n_s$, where $1 \leq n_s < n$. Since A_s is fully indecomposable, it follows by the induction hypothesis that

$$\operatorname{per}(A_s) \geq \sigma(A_s) - 2n_s + 2.$$

But $n_s = n - (s-1)$, and $\sigma(A_s) = \sigma(B) - s - (s-1) = \sigma(A) - 2s + 1$, and therefore

$$\begin{aligned} \operatorname{per}(A_s) &\geq \sigma(A) - 2s + 1 - 2(n - s + 1) + 2 \\ &= \sigma(A) - 2n + 1. \end{aligned} \tag{4}$$

Let the 1 in E_1 be in the $(1, j)$ position in B. Expanding the permanent of B by the first row we get

$$\begin{aligned} \operatorname{per}(B) &= \operatorname{per}(B(1|1)) + \operatorname{per}(B(1|j)) \\ &\geq \operatorname{per}(A_s) + 1, \end{aligned}$$

by Theorem 4.1. Hence

$$\begin{aligned} \operatorname{per}(A) &= \operatorname{per}(B) \\ &\geq \operatorname{per}(A_s) + 1 \\ &\geq \sigma(A) - 2n + 2, \end{aligned}$$

by (4).

Now, suppose that A is fully indecomposable but not nearly decomposable. Then there exists an entry $a_{i_1 j_1}$ in A so that $A - E_{i_1 j_1}$ is a fully indecomposable (0, 1)-matrix. If $A - E_{i_1 j_1}$ is not nearly decomposable, then there must exist an entry $a_{i_2 j_2} = 1$ in $A - E_{i_1 j_1}$ such that $A - E_{i_1 j_1} - E_{i_2 j_2}$ is fully indecomposable, and so on. Thus we must finally obtain a nearly decomposable (0, 1)-matrix C satisfying

$$A = C + \sum_{t=1}^{m} E_{i_t j_t}.$$

By Corollary 4.3,

$$\operatorname{per}(A) \geq \operatorname{per}(C) + m,$$

and applying inequality (3) to the nearly decomposable (0, 1)-matrix C, we obtain

$$\begin{aligned} \operatorname{per}(A) &\geq \sigma(C) - 2n + 2 + m \\ &= \sigma(A) - 2n + 2, \end{aligned}$$

since $\sigma(A) = \sigma(C) + m$. ■

It can be shown that the lower bound in Theorem 6.3 is the best possible, in the sense that for each $n \geq 3$ and each N, $2n \leq N \leq 3(n - 1)$, there exists a nearly decomposable $n \times n$ (0, 1)-matrix A such that $\sigma(A) = N$, and $\operatorname{per}(A) = \sigma(A) - 2n + 2$ [26]. However, if more information about the matrix is available the bound in Theorem 6.3 can be improved. Gibson [10] used Hall's inequality in Theorem 6.2 to obtain the following improvement of Minc's inequality in Theorem 6.3.

Theorem 6.4. *If A is a fully indecomposable $n \times n$ matrix with at least t ones in each row, then*

$$\operatorname{per}(A) \geq \sigma(A) - 2n + 2 + \sum_{i=1}^{t-1} (i! - 1). \tag{5}$$

Proof. Use induction on t. If $t = 1$ or 2, then inequality (5) reduces to inequality (3). Assume that $t \geq 3$ and that (5) holds for all $k < t$. Since each row of A has at least t ones, it follows from Theorem 5.2 that A cannot be nearly decomposable. Thus there must exist a position (p, q) such that $a_{pq} = 1$, and $B = A - E_{pq}$ is fully indecomposable and has at least $t - 1$ ones in each of its rows. By the induction hypothesis,

$$\operatorname{per}(B) \geq \sigma(A) - 2n + 1 + \sum_{i=1}^{t-2} (i! - 1). \tag{6}$$

Now, $\operatorname{per}(A) = \operatorname{per}(B) + \operatorname{per}(A(p|q))$. Since A is fully indecomposable, the permanent of $A(p|q)$ is positive, and each row sum of $A(p|q)$ is at least $t-1$. Hence, by Theorem 6.2,

$$\operatorname{per}(A(p|q)) \geq (t-1)!. \tag{7}$$

The result follows from (6) and (7). ■

We conclude this section with an upper bound for the permanents of (0, 1)-matrices. The bound was conjectured and proved in special cases by Minc [21] and proved completely by Brègman [3]. We give below an elegant proof due to Schrijver [27]. We start with two preliminary results. Let r_i denote the ith row sum of an $n \times n$ matrix $A = (a_{ij})$, that is, $r_i = \sum_{i=1}^{n} a_{ij}$, $i = 1, 2, \ldots, n$.

Lemma 6.1. *If $t_1, t_2, \ldots, t_n$ are nonnegative real numbers, then*

$$\left(\frac{1}{n}\sum_{k=1}^{n} t_k\right)^{\Sigma t_k} \leq \prod_{k=1}^{n} t_k^{t_k}, \tag{8}$$

where the summation in the exponent on the left-hand side extends from 1 *to* n, *and* 0^0 *denotes* 1.

The lemma is an immediate consequence of the convexity of the function $x \log x$. For,

$$\left(\frac{1}{n}\sum_{k=1}^{n} t_k\right)\log\left(\frac{1}{n}\sum_{k=1}^{n} t_k\right) \leq \frac{1}{n}\sum_{k=1}^{n} t_k \log t_k,$$

which after multiplying by n and taking exponents of both sides, yields (8).

Lemma 6.2. *Let $A = (a_{ij})$ be an $n \times n$ (0, 1)-matrix with positive permanent, and let S be the set of permutations corresponding to positive diagonals of A, that is, $\sigma \in S$ if and only if $\prod_{i=1}^{n} a_{i,\sigma i} = 1$. Then*

$$\prod_{i=1}^{n} \prod_{\substack{k \\ a_{ik}=1}} (\operatorname{per}(A(i|k)))^{\operatorname{per}(A(i|k))} = \prod_{\sigma \in S} \prod_{i=1}^{n} \operatorname{per}(A(i|\sigma i)), \tag{9}$$

and

$$\prod_{i=1}^{n} r_i^{\operatorname{per} A} = \prod_{\sigma \in S} \prod_{i=1}^{n} r_i. \tag{10}$$

Proof. For a given i and k, the number of factors $\operatorname{per}(A(i|k))$ on the left-hand side of (9) is $\operatorname{per}(A(i|k))$ if $a_{ik} = 1$, and zero otherwise. The number of such factors on the right-hand side of (9) is equal to the number of permutations σ in S satisfying $\sigma i = k$, which is $\operatorname{per}(A(i|k))$ or zero, according as $a_{ik} = 1$ or 0.

It is easy to see that for a given i the number of factors r_i on both sides of (10) is $\operatorname{per}(A)$. ■

Theorem 6.5. *Let* $A = (a_{ij})$ *be an* $n \times n$ $(0,1)$*-matrix with row sums* $r_1, r_2, \ldots, r_n$. *Then*

$$\operatorname{per}(A) \le \prod_{i=1}^{n} r_i!^{1/r_i}. \tag{11}$$

Proof (Schrijver [26]). Use induction on n. By Lemma 6.1,

$$\begin{aligned}(\operatorname{per} A)^{n \operatorname{per} A} &= \prod_{i=1}^{n} (\operatorname{per} A)^{\operatorname{per} A} \\ &= \prod_{i=1}^{n} \left(\sum_{k=1}^{n} a_{ik} \operatorname{per} A(i|k) \right)^{\sum a_{ik} \operatorname{per} A(i|k)} \\ &\le \prod_{i=1}^{n} \left(r_i^{\operatorname{per} A} \prod_{\substack{k \\ a_{ik}=1}} \operatorname{per} A(i|k)^{\operatorname{per} A(i|k)} \right),\end{aligned}$$

and thus, by Lemma 6.2,

$$(\operatorname{per} A)^{n \operatorname{per} A} \le \prod_{\sigma \in S} \left(\left(\prod_{i=1}^{n} r_i \right) \left(\prod_{i=1}^{n} \operatorname{per} A(i|\sigma i) \right) \right).$$

We now apply the induction hypothesis to each $A(i|\sigma i)$:

$$\begin{aligned}\prod_{i=1}^{n} \operatorname{per} A(i|\sigma i) &\le \prod_{i=1}^{n} \left(\prod_{\substack{j \ne i \\ a_{j,\sigma i}=0}} r_j!^{1/r_j} \right) \left(\prod_{\substack{j \ne i \\ a_{j,\sigma i}=1}} (r_j - 1)!^{1/(r_j-1)} \right) \\ &= \prod_{j=1}^{n} \left(\prod_{\substack{i \ne j \\ a_{j,\sigma i}=0}} r_j!^{1/r_j} \right) \left(\prod_{\substack{i \ne j \\ a_{j,\sigma i}=1}} (r_j - 1)!^{1/(r_j-1)} \right) \\ &= \prod_{j=1}^{n} r_j!^{(n-r_j)/r_j} (r_j - 1)!^{(r_j-1)/(r_j-1)}.\end{aligned}$$

The first equality is just a result of a change in the order of multiplication, and the second equality is obtained by counting the number of factors $r_j!^{1/r_j}$ and factors $(r_j - 1)!^{1/(r_j-1)}$. Clearly, for fixed σ and j, the number of i satisfying $i \ne j$ and $a_{j,\sigma i} = 0$ is $n - r_j$, and the number of i satisfying $i \ne j$ and

$a_{j,\sigma i} = 1$ is $r_j - 1$ (since $a_{j,\sigma j} = 1$). Hence

$$\begin{aligned}(\operatorname{per} A)^{n \operatorname{per} A} &\le \prod_{\sigma \in S}\left(\left(\prod_{i=1}^{n} r_i\right)\left(\prod_{j=1}^{n} r_j!^{(n-r_j)/r_j}(r_j - 1)!\right)\right)\\ &= \prod_{\sigma \in S}\left(\prod_{i=1}^{n} r_i!^{n/r_i}\right)\\ &= \left(\prod_{i=1}^{n} r_i!^{1/r_i}\right)^{n \operatorname{per} A},\end{aligned}$$

and the result follows. ∎

PROBLEMS

1 Let $S = \{s_1, s_2, s_3, s_4, s_5\}$ and $S_1 = \{s_2, s_5\}$, $S_2 = \{s_1, s_4, s_5\}$, $S_3 = \{s_1, s_4, s_5\}$, $S_4 = \{s_3, s_4\}$.

(a) Find all the SDRs of the above configuration.

(b) Find the permanent of the incidence matrix of the above configuration using the Laplace expansion (Theorem 1.2) on the first row of the matrix.

2 Prove that the product of fully indecomposable nonnegative matrices is fully indecomposable ([17]).

3 Call an $m \times n$ nonnegative matrix A, $m \le n$, *fully indecomposable* if $\operatorname{Per}(A(i|j)) > 0$ for $i = 1, 2, \ldots, m$, $j = 1, 2, \ldots, n$. Interpret this definition in the context of incidence matrices for configurations of subsets. What does the definition say about the $m \times m$ submatrices of A?

4 Extend the definition of near decomposability to $m \times n$ nonnegative matrices, $m \le n$. Give an example of a 3×4 nearly decomposable matrix.

5 Call an $n \times n$ nonnegative matrix A *k-indecomposable* if $\operatorname{per}(A(\alpha|\beta)) > 0$ for all α and β in $Q_{k,n}$. Show that if A is fully indecomposable (i.e., 1-indecomposable), then AA^{T} is 2-indecomposable.

6 Show that if A is fully indecomposable, then AA^{T} and $A^{\mathrm{T}}A$ are fully indecomposable. Is the converse true?

7 Show that if A is fully indecomposable, then A is primitive. Is the converse true?

8 Let D be the directed graph

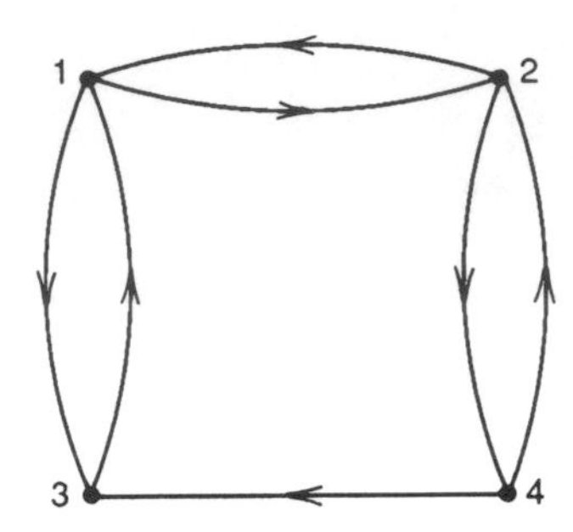

(a) Use the adjacency matrix of D to find all pairs of vertices connected by paths (i) of length 2, (ii) of length 3, (iii) of length 4.

(b) Show that every cycle in D is of even length.

9 Find the eigenvalues of matrices B_n and C_n in Lemma 5.5.

10 Let A be the 4×4 matrix

$$A = \left[\begin{array}{c|c} 1 & F_2 \\ \hline F_1 & C_3 \end{array}\right],$$

where $F_1 = [1 \quad 0 \quad 0]^{\mathrm{T}}$, $F_2 = [1 \quad 0 \quad 0]$, and C_n is the matrix defined in Lemma 5.5(b). Find permutation matrices P and Q such that $PAQ = C_4$.

11 Construct a doubly stochastic matrix with the same zero patterns as C_4.

12 Show that the g.c.d. of the lengths of all circuits in a directed graph is equal to the index of imprimitivity of the graph.

13 Use graph methods to determine which of the following matrices are irreducible:

$$A_1 = \begin{bmatrix} 0 & 1 & 0 & 1 \\ 1 & 0 & 1 & 0 \\ 0 & 1 & 0 & 0 \\ 1 & 0 & 0 & 0 \end{bmatrix}, \qquad A_2 = \begin{bmatrix} 1 & 0 & 0 & 1 \\ 1 & 0 & 1 & 0 \\ 0 & 1 & 0 & 1 \\ 1 & 0 & 0 & 1 \end{bmatrix},$$

$$A_3 = \begin{bmatrix} 0 & 1 & 0 & 1 \\ 0 & 0 & 1 & 0 \\ 0 & 0 & 0 & 1 \\ 1 & 0 & 0 & 0 \end{bmatrix}, \qquad A_4 = \begin{bmatrix} 0 & 1 & 1 & 0 \\ 0 & 0 & 1 & 0 \\ 0 & 0 & 0 & 1 \\ 1 & 0 & 0 & 0 \end{bmatrix},$$

$$A_5 = \begin{bmatrix} 0 & 1 & 0 & 0 & 1 & 0 \\ 0 & 0 & 0 & 0 & 0 & 1 \\ 0 & 1 & 0 & 0 & 0 & 0 \\ 1 & 0 & 1 & 0 & 0 & 0 \\ 0 & 0 & 0 & 1 & 0 & 1 \\ 0 & 1 & 0 & 0 & 0 & 0 \end{bmatrix}, \qquad A_6 = \begin{bmatrix} 0 & 1 & 0 & 0 & 1 & 0 \\ 0 & 0 & 0 & 0 & 0 & 1 \\ 0 & 1 & 0 & 0 & 0 & 0 \\ 0 & 0 & 1 & 0 & 0 & 0 \\ 0 & 0 & 0 & 1 & 0 & 1 \\ 1 & 0 & 0 & 0 & 0 & 0 \end{bmatrix}.$$

14 **(a)** Which of the matrices in Problem 13 are primitive?

(b) Determine the index of imprimitivity for the irreducible matrices in Problem 13.

(c) Which of the matrices in Problem 13 are nearly decomposable?

15 Construct a primitive matrix A (other than the matrix given as an example after the proof of Theorem 3.4) such that $A * A^2 = 0$.

16 Show by a counterexample that the converse of Theorem 4.3 is not true.

17 Let D be a strongly connected directed graph with index of imprimitivity h. Prove that the lengths of all paths connecting two fixed vertices are congruent modulo h.

18 Let A be the adjacency matrix of a directed graph D. Let S be a subgraph of D consisting of disjoint circuits (i.e., circuits which have no vertex in common). Show that the entries in A corresponding to arcs of S form a positive diagonal in a principal submatrix of A.

19 Let A be a nonnegative $n \times n$ matrix. Suppose that the coefficient of λ^{n-k} in the characteristic polynomial of A is nonzero. Show that the directed graph associated with A contains a subgraph consisting of disjoint circuits, the sum of whose lengths is k.

20 Let $\lambda^n + a_1\lambda^{n_1} + a_2\lambda^{n_2} + \cdots + a_m\lambda^{n_m}$, where $n > n_1 > n_2 > \cdots > n_m$ and $a_t \neq 0$, $t = 1, 2, \ldots, m$, be the characteristic polynomial of an irreducible matrix with index of imprimitivity h. Use the results in Problems 12 and 19 to show that h divides $n - n_t$, $t = 1, 2, \ldots, m$ (cf. Theorem 1.3, Chapter III).

21 Prove Lemma 5.1.

22 Show that the converse of Lemma 5.2 is not true.

23 Prove Theorem 5.3.

24 Construct nearly decomposable 8×8 (0,1)-matrices with N positive entries, for $N = 16, 17, 18, 19, 20$, and 21.

25 Construct nearly irreducible 8×8 (0,1)-matrices with N positive entries, for $N = 8, 9, 10, 11, 12, 13$, and 14.

26 Show by an example that a fully indecomposable nonnegative $n \times n$ matrix with exactly $2n + 1$ positive entries is not necessarily nearly decomposable.

27 A (*not directed*) *graph* G is defined as a nonempty set V of vertices together with a set E containing *unordered* pairs of vertices of V called *edges* of G. Graph G is said to be *bipartite* if V can be partitioned into two subsets, M and N, so that each edge of G has one of its vertices in M and the other in N.

Describe the adjacency matrix of a bipartite graph.

28 Let G be a bipartite graph with m vertices in M and n vertices in N, $m \leq n$. A *perfect matching* of G is a set of m edges such that no two edges in the set have a common vertex. Find all perfect matchings of the bipartite graph

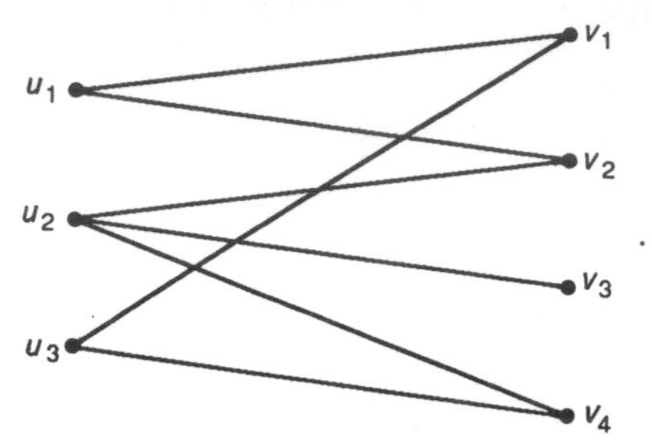

29 Let G be a bipartite graph with $M = \{u_1, u_2, \ldots, u_m\}$ and $N = \{v_1, v_2, \ldots, v_n\}$, $m \leq n$. The *incidence matrix* of G is the $m \times n$ (0,1)-matrix $A = (a_{ij})$, where $a_{ij} = 1$ or 0 according as (u_i, v_j) is an edge of G or not.

(a) Construct the incidence matrix of the bipartite graph in Problem 28.

(b) Construct the bipartite graph whose incidence matrix is

$$\begin{bmatrix} 0 & 1 & 0 & 1 & 0 \\ 1 & 1 & 1 & 0 & 1 \\ 1 & 0 & 1 & 1 & 0 \end{bmatrix}.$$

(c) What in the incidence matrix of a bipartite graph G corresponds to a perfect matching of G?

30 How can the number of distinct perfect matchings in a bipartite graph be determined from its incidence matrix? Compute this number for the graph in Problem 28 using the incidence matrix of the graph.

31 Compute the lower bounds given in Theorems 6.2–6.4 and the upper bound given in Theorem 6.5 for the permanents of the following matrices:

$$A = \begin{bmatrix} 1 & 1 & 1 & 0 & 0 & 0 \\ 0 & 1 & 1 & 1 & 0 & 0 \\ 0 & 0 & 1 & 1 & 1 & 0 \\ 0 & 0 & 0 & 1 & 1 & 1 \\ 1 & 0 & 0 & 0 & 1 & 1 \\ 1 & 1 & 0 & 0 & 0 & 1 \end{bmatrix}, \qquad B = \begin{bmatrix} 1 & 1 & 0 & 0 & 0 & 0 \\ 1 & 1 & 1 & 0 & 0 & 0 \\ 0 & 1 & 1 & 1 & 0 & 0 \\ 0 & 0 & 1 & 1 & 1 & 0 \\ 0 & 0 & 0 & 1 & 1 & 1 \\ 0 & 0 & 0 & 0 & 1 & 1 \end{bmatrix},$$

$$C = \begin{bmatrix} 1 & 1 & 1 & 1 & 0 & 0 \\ 0 & 1 & 1 & 1 & 1 & 0 \\ 0 & 0 & 1 & 1 & 1 & 1 \\ 1 & 0 & 0 & 1 & 1 & 1 \\ 1 & 1 & 0 & 0 & 1 & 1 \\ 1 & 1 & 1 & 0 & 0 & 1 \end{bmatrix}, \qquad D = \begin{bmatrix} 1 & 1 & 0 & 0 & 0 & 0 \\ 1 & 1 & 1 & 0 & 0 & 0 \\ 1 & 1 & 1 & 1 & 0 & 0 \\ 1 & 1 & 1 & 1 & 1 & 0 \\ 1 & 1 & 1 & 1 & 1 & 1 \\ 1 & 1 & 1 & 1 & 1 & 1 \end{bmatrix}.$$

[The exact values are per(A) = 20, per(B) = 13, per(C) = 80, per(D) = 32.]

32 Prove the case $t > m$ of Theorem 6.2 by bordering matrix A by a $(t - m) \times n$ matrix all of whose entries are 1 and applying the other part of the theorem to the resulting $t \times n$ (0, 1)-matrix.

33 Show that the condition Per(A) > 0 in the statement of Theorem 6.2 is essential.

34 Verify that Theorem 6.3 holds for nearly decomposable $n \times n$ (0, 1)-matrices for $n = 1$, 2, and 3.

REFERENCES

1. R. Aharoni, On a theorem of Dénes König, *Linear and Multilinear Algebra* **4** (1976), 31–32.
2. C. Berge, *The Theory of Graphs*, Methuen, London, 1962.
3. L. M. Brègman, Certain properties of nonnegative matrices and their permanents, *Dokl. Akad. Nauk SSSR* **211** (1973), 27–30 (in Russian). Translated in *Soviet Math. Dokl.* **14** (1973), 945–949.
4. R. Brualdi, S. Parter, and H. Schneider, The diagonal equivalence of a nonnegative matrix to a stochastic matrix, *J. Math. Anal. Appl.* **16** (1966), 31–50.
5. N. G. de Bruijn and P. Erdös, On a combinatorial problem, *Indag. Math.* **10** (1948), 421–423.
6. A. L. Dulmage and N. S. Mendelsohn, Graphs and matrices, *Graph Theory and Theoretical Physics* (F. Harary, ed.), Academic Press, London, 1967, 167–227.
7. G. Edgar, Proof (unpublished).
8. G. Frobenius, Über Matrizen aus nicht negativen Elementen, *S.-B. K. Preuss. Akad. Wiss. Berlin* (1912), 456–477.
9. G. Frobenius, Über zerlegbare Determinanten, *S.-B. K. Preuss. Akad. Wiss. Berlin* (1917), 274–277.
10. P. M. Gibson, A lower bound for the permanent of a (0,1)-matrix, *Proc. Amer. Math. Soc.* **33** (1972), 245–246.
11. M. Hall, Jr., Distinct representatives of subsets, *Bull. Amer. Math. Soc.* **54** (1948), 922–926.
12. F. Harary, Determinants, permanents and bipartite graphs, *Math. Mag.* **42** (1969), 146–148.
13. D. J. Hartfiel, A simplified form for nearly reducible and nearly decomposable matrices, *Proc. Amer. Math. Soc.* **24** (1970), 388–393.
14. J. E. Hopcroft and R. M. Karp, An $n^{5/2}$ algorithm for maximum matchings in bipartite graphs, *SIAM J. Comput.* **2** (1973), 225–231.
15. D. König, Vonalrendszerek és determinánsok (Graphs and determinants), *Mat. Természettud. Ertesítö* **33** (1915), 221–229.
16. D. König, *Theorie der endlichen und unendlichen Graphen*, Akademische Verlagsgesellschaft, Leipzig, 1936.
17. M. Lewin, On nonnegative matrices, *Pacific J. Math.* **36** (1971), 753–759.
18. M. Lewin, On exponents of primitive matrices, *Numer. Math.* **18** (1971/72), 154–161.
19. H. B. Mann and H. J. Ryser, Systems of distinct representations, *Amer. Math. Monthly* **60** (1953), 397–401.
20. M. Marcus and H. Minc, Disjoint pairs of sets and incidence matrices, *Illinois J. Math.* **7** (1963), 137–147.

21. H. Minc, Upper bounds for permanents of (0,1)-matrices, *Bull. Amer. Math. Soc.* **69** (1963), 789–791.

22. H. Minc, On lower bounds for permanents of (0,1) matrices, *Proc. Amer. Math. Soc.* **22** (1969), 117–123.

23. H. Minc, Nearly decomposable matrices, *Linear Algebra Appl.* **5** (1972), 181–187.

24. H. Minc, A remark on a theorem of M. Hall, *Canad. Math. Bull.* **17** (1974), 547–548.

25. H. Minc, *Permanents*, *Encyclopedia of Mathematics and Its Applications*, vol. 6, Addison-Wesley, Reading, Mass., 1978.

26. E. J. Roberts, The fully indecomposable matrix and its associated bipartite graph—an investigation of combinatorial and structural properties (Ph.D. dissertation, University of Houston, Texas, NASA TM X-58037, 1970.

27. A. Schrijver, A short proof of Minc's conjecture, *J. Combin. Theory Ser. A* **25** (1978), 80–81.

28. R. Sinkhorn and P. Knopp, Problems involving diagonal products in nonnegative matrices, *Trans. Amer. Math. Soc.* **136** (1969), 67–75.

29. R. S. Varga, *Matrix Iterative Analysis*, Prentice-Hall, Englewood Cliffs, N.J., 1962.

V

Doubly Stochastic Matrices

5.1. DEFINITIONS AND EARLY RESULTS

In this chapter we study doubly stochastic matrices, a class of nonnegative matrices with important applications in many areas of mathematics and the physical sciences: linear algebra, theory of inequalities, combinatorial matrix theory, combinatorics, probability, physical chemistry, etc.

Definition 1.1. A real matrix is said to be *doubly quasi-stochastic* if each of its row and column sums is 1. A nonnegative doubly quasi-stochastic matrix is called *doubly stochastic*. The set of $n \times n$ doubly stochastic matrices is denoted by Ω_n.

Clearly, a doubly quasi-stochastic matrix must be square. It follows from the definition that an $n \times n$ matrix A is doubly quasi-stochastic if and only if 1 is an eigenvalue of A, and $(1, 1, \ldots, 1)$ is an eigenvector corresponding to this eigenvalue for both A and A^{T}. Thus a nonnegative $n \times n$ matrix A is doubly stochastic if and only if

$$AJ_n = J_n A = J_n,$$

where J_n is the $n \times n$ matrix all of whose entries are $1/n$.

We begin with two interesting early results due to König [8] and Schur [23].

Theorem 1.1 (König [8]). *Every doubly stochastic matrix has a positive diagonal.*

Proof. If a matrix A in Ω_n had no positive diagonals, then the permanent of A would vanish and, by the Frobenius–König theorem, there would exist permutation matrices P and Q such that

$$PAQ = \begin{bmatrix} B & C \\ 0 & D \end{bmatrix},$$

where the zero block in the lower left corner is $p \times q$, with $p + q = n + 1$.

Let $\sigma(X)$ denote the sum of the entries in matrix X. Then

$$\begin{aligned} n &= \sigma(PAQ) \\ &\geq \sigma(B) + \sigma(D) \\ &= q + p \\ &= n + 1. \end{aligned}$$

This contradiction proves the theorem. ■

Corollary 1.1. *The permanent of a doubly stochastic matrix is positive.*

Theorem 1.2 (Schur [23]). *Let $H = (h_{ij})$ be a hermitian $n \times n$ matrix with eigenvalues $\lambda_1, \lambda_2, \ldots, \lambda_n$, and let $h = [h_{11}, h_{22}, \ldots, h_{nn}]^{\mathrm{T}}$ and $\lambda = [\lambda_1, \lambda_2, \ldots, \lambda_n]^{\mathrm{T}}$. Then there exists a doubly stochastic matrix S such that $h = S\lambda$.*

Proof. Let $U = (u_{ij})$ be a unitary matrix such that

$$H = U\operatorname{diag}(\lambda_1, \lambda_2, \ldots, \lambda_n)U^*.$$

Then

$$\begin{aligned} h_{ii} &= \sum_{t=1}^{n} u_{it}\lambda_t \bar{u}_{it} \\ &= \sum_{t=1}^{n} |u_{it}|^2 \lambda_t \\ &= \sum_{t=1}^{n} s_{it}\lambda_t, \qquad i = 1, 2, \ldots, n, \end{aligned}$$

where $s_{it} = |u_{it}|^2$, $i, t = 1, 2, \ldots, n$. Clearly, the $n \times n$ matrix $S = (s_{ij})$ is doubly stochastic. The result follows. ■

Definition 1.2. (a) An $n \times n$ matrix $A = (a_{ij})$ is called *orthostochastic* if there exists a (real) orthogonal matrix $T = (t_{ij})$ such that $a_{ij} = t_{ij}^2$, $i, j = 1, 2, \ldots, n$.

(b) An $n \times n$ matrix $A = (a_{ij})$ is called *Schur-stochastic* (or *unitary-stochastic*) if there exists a unitary matrix $U = (u_{ij})$ such that $a_{ij} = |u_{ij}|^2$ for all i and j.

The matrix S in Theorem 1.2 is Schur-stochastic. Clearly, every orthostochastic matrix is Schur-stochastic, and every Schur-stochastic matrix is doubly stochastic. However, not every doubly stochastic matrix is Schur-stochastic, nor is every Schur-stochastic matrix orthostochastic.

Example 1.1. (a) The doubly stochastic matrix

$$A = (a_{ij}) = \frac{1}{2}\begin{bmatrix} 0 & 1 & 1 \\ 1 & 0 & 1 \\ 1 & 1 & 0 \end{bmatrix}$$

is not Schur-stochastic. For, if $U = (u_{ij})$ is any 3×3 matrix such that $a_{ij} = |u_{ij}|^2$, $i, j = 1, 2, 3$, then $u_{11} = u_{22} = u_{33} = 0$, but $u_{11}\bar{u}_{21} + u_{12}\bar{u}_{22} + u_{13}\bar{u}_{23} = u_{13}\bar{u}_{23} \neq 0$ since the moduli of u_{13} and $\bar{u}_{23}$ are both $1/\sqrt{2}$. Thus U cannot be unitary, and A is not Schur-stochastic.

(b) The doubly stochastic matrix J_3 is Schur-stochastic. For, if $U = (u_{ij})$ is the unitary matrix

$$\frac{1}{\sqrt{3}}\begin{bmatrix} 1 & \theta & \theta^2 \\ \theta^2 & 1 & \theta \\ \theta & \theta^2 & 1 \end{bmatrix},$$

where θ is a primitive cube root of 1, then $|u_{ij}|^2 = \frac{1}{3}$ for all i and j. However, J_3 is not orthostochastic. For, if $T = (t_{ij})$ were a real 3×3 matrix such that $t_{ij}^2 = \frac{1}{3}$ for all i and j, then $t_{11}t_{21} + t_{12}t_{22} + t_{13}t_{23}$ could not vanish (it would be equal to -1, or $-\frac{1}{3}$, or $\frac{1}{3}$, or 1), and therefore T could not be orthogonal. ■

We note the following property of doubly stochastic matrices.

Lemma 1.1. *A product of doubly stochastic matrices is doubly stochastic.*

For, if A and B are doubly stochastic $n \times n$ matrices (and therefore $AJ_n = J_nA = BJ_n = J_nB = J_n$), then their product is nonnegative,

$$(AB)J_n = A(BJ_n) = AJ_n = J_n,$$

and

$$J_n(AB) = (J_nA)B = J_nB = J_n,$$

and therefore AB is doubly stochastic. It is easily seen that the same is true of a product of any number of doubly stochastic matrices.

Definition 1.3. A doubly stochastic $n \times n$ matrix with $n - 2$ main diagonal entries equal to 1 is called an *elementary* doubly stochastic matrix. In other words, $A = (a_{ij}) \in \Omega_n$ is elementary if $a_{ss} = a_{tt} = 1 - \theta$, $a_{st} = a_{ts} = \theta$, for some integers s, t, $1 \leq s < t \leq n$, and a real number θ, $0 \leq \theta \leq 1$, and $a_{ij} = \delta_{ij}$ otherwise.

It follows from Lemma 1.1 that a product of elementary doubly stochastic matrices is doubly stochastic. However, the converse is not true: Not every doubly stochastic matrix can be expressed as a product of elementary doubly stochastic matrices. For example, the doubly stochastic matrix A in Example 1.1(a) is neither an elementary doubly stochastic matrix nor is it a product of such matrices (see Problem 7).

Reducible and imprimitive irreducible doubly stochastic matrices have special structural properties.

Theorem 1.3. *A reducible doubly stochastic matrix is cogredient to a direct sum of doubly stochastic matrices.*

Proof. Let A be a reducible $n \times n$ doubly stochastic matrix. Then A is cogredient to a matrix of the form

$$B = \begin{bmatrix} X & Y \\ 0 & Z \end{bmatrix},$$

where X is k-square and Z is $(n-k)$-square. Clearly, B is doubly stochastic. The sum of the entries in the first k columns of B is k, and all nonzero entries in these columns are contained in X. Therefore

$$\sigma(X) = k.$$

Similarly, by considering the last $n-k$ rows of B, we can conclude that

$$\sigma(Z) = n - k.$$

But

$$\begin{aligned} n &= \sigma(B) \\ &= \sigma(X) + \sigma(Y) + \sigma(Z) \\ &= k + \sigma(Y) + (n-k) \\ &= n + \sigma(Y), \end{aligned}$$

and therefore

$$\sigma(Y) = 0.$$

Hence

$$Y = 0,$$

and A is cogredient to $B = X \dotplus Z$, where X and Z are clearly doubly stochastic. ■

If in the above proof either X or Z happens to be reducible, then it is also cogredient to a direct sum of doubly stochastic matrices. We have therefore the following results.

Corollary 1.2. *A reducible doubly stochastic matrix is cogredient to a direct sum of irreducible doubly stochastic matrices.*

Corollary 1.3. *The elementary divisors corresponding to* 1, *the maximal eigenvalue of a doubly stochastic matrix, are linear.*

The following analogous results for partly decomposable doubly stochastic matrices can be proved similarly.

Corollary 1.4. *A partly decomposable doubly stochastic matrix is p-equivalent to a direct sum of doubly stochastic matrices.*

Corollary 1.5. *A partly decomposable doubly stochastic matrix is p-equivalent to a direct sum of fully indecomposable doubly stochastic matrices.*

Our next theorem, due to Marcus, Minc, and Moyls [12], describes the structure of imprimitive irreducible doubly stochastic matrices.

Theorem 1.4. *Let A be an irreducible doubly stochastic $n \times n$ matrix with index of imprimitivity h. Then h divides n, and the matrix A is cogredient to a matrix in the superdiagonal block form*

$$\begin{bmatrix} 0 & A_{12} & 0 & \cdots & 0 & 0 \\ 0 & 0 & A_{23} & \cdots & 0 & 0 \\ \vdots & & & \ddots & & \vdots \\ 0 & 0 & & \cdots & 0 & A_{h-1,h} \\ A_{h1} & 0 & & \cdots & 0 & 0 \end{bmatrix}, \tag{1}$$

where all the blocks are (n/h)-square.

Proof. By Theorem 3.1, Chapter III, A is cogredient to a partitioned matrix of the form (1), where the zero blocks along the main diagonal are square. Clearly, each of the blocks $A_{12}, A_{23}, \ldots, A_{h-1,h}, A_{h1}$ must be doubly stochastic and therefore square. But this implies that the zero blocks along the main diagonal are of the same order. The result now follows. ■

Corollary 1.6. *A doubly stochastic matrix is p-equivalent to a direct sum of primitive matrices.*

5.2. THEOREMS OF MUIRHEAD AND OF HARDY, LITTLEWOOD, AND PÓLYA

We introduce the following notation. If $\gamma = (\gamma_1, \gamma_2, \dots, \gamma_n)$ is a real n-tuple, then $\gamma^* = (\gamma_1^*, \gamma_2^*, \dots, \gamma_n^*)$ denotes the n-tuple γ rearranged in nonincreasing order, $\gamma_1^* \geq \gamma_2^* \geq \cdots \geq \gamma_n^*$.

Definition 2.1. A nonnegative n-tuple $\alpha = (\alpha_1, \alpha_2, \dots, \alpha_n)$ is said to be *majorized* by a nonnegative n-tuple $\beta = (\beta_1, \beta_2, \dots, \beta_n)$ if

$$\alpha_1^* + \alpha_2^* + \cdots + \alpha_k^* \leq \beta_1^* + \beta_2^* + \cdots + \beta_k^*,$$

for $k = 1, 2, \dots, n-1$, and

$$\alpha_1 + \alpha_2 + \cdots + \alpha_n = \beta_1 + \beta_2 + \cdots + \beta_n.$$

This is denoted by $\alpha \prec \beta$.

One of the most important and elegant results in the area of majorization of nonnegative n-tuples, with numerous applications in many areas of mathematics, is due to Muirhead [22].

Theorem 2.1. *Let $c = (c_1, c_2, \dots, c_n)$ be a positive n-tuple, and let $\alpha = (\alpha_1, \alpha_2, \dots, \alpha_n)$ and $\beta = (\beta_1, \beta_2, \dots, \beta_n)$ be n-tuples of nonnegative integers. Let $A(c)$ and $B(c)$ be $n \times n$ matrices whose (i, j) entries are $c_i^{\alpha_j}$ and $c_i^{\beta_j}$, respectively. Then*

$$\alpha \prec \beta,$$

if and only if

$$\operatorname{per}(A(c)) \leq \operatorname{per}(B(c)),$$

for all positive n-tuples c.

The proof of Theorem 2.1 is postponed to the end of the section.

Hardy, Littlewood, and Pólya [6] extended Muirhead's theorem to any nonnegative n-tuples α and β, and proved the following result.

Theorem 2.2. *Let α and β be real nonnegative n-tuples. Then $\alpha \prec \beta$ if and only if there exists a doubly stochastic $n \times n$ matrix S such that*

$$\alpha = S\beta.$$

We first prove the following lemmas.

Lemma 2.1. *If $\alpha = (\alpha_1, \alpha_2, \ldots, \alpha_n)$ and $\beta = (\beta_1, \beta_2, \ldots, \beta_n)$ are nonnegative n-tuples, then*

$$\sum_{i=1}^{n} \alpha_i \beta_i \le \sum_{i=1}^{n} \alpha_i^* \beta_i^*. \tag{1}$$

Proof. We can assume without loss of generality that $\alpha = \alpha^*$. Suppose that $\beta_s < \beta_t$ for some $s < t$. Then

$$(\alpha_s \beta_t + \alpha_t \beta_s) - (\alpha_s \beta_s + \alpha_t \beta_t) = (\alpha_s - \alpha_t)(\beta_t - \beta_s) \ge 0.$$

In other words, the sum on the left-hand side of (1) is not diminished if β_s and β_t are transposed. A finite number of such transpositions will produce the sum on the right-hand side of (1). ∎

Lemma 2.2. *Let k and n be positive integers, $k \le n$, and let $c_1, c_2, \ldots, c_n, d_1, d_2, \ldots, d_n$ be nonnegative numbers such that $c_i \le 1$, $i = 1, 2, \ldots, n$, $\sum_{i=1}^{n} c_i = k$, and $d_1 \ge d_2 \ge \cdots \ge d_n \ge 0$. Then*

$$\sum_{i=1}^{n} c_i d_i \le \sum_{i=1}^{k} d_i.$$

The lemma is nearly obvious intuitively. However, we give here a formal proof.

Proof. We have

$$\begin{aligned}
\sum_{i=1}^{k} d_i - \sum_{i=1}^{n} c_i d_i &= \sum_{i=1}^{k} (1 - c_i) d_i - \sum_{i=k+1}^{n} c_i d_i \\
&\ge \sum_{i=1}^{k} (1 - c_i) d_k - \sum_{i=k+1}^{n} c_i d_k \\
&= d_k \left(k - \sum_{i=1}^{k} c_i \right) - d_k \sum_{i=k+1}^{n} c_i \\
&= 0. \quad ∎
\end{aligned}$$

Proof of Theorem 2.2. Let $\alpha = S\beta$. We can assume without loss of generality that $\alpha = \alpha^*$. Let k be any integer, $1 \le k < n$. Then

$$\begin{aligned}
\sum_{i=1}^{k} \alpha_i^* &= \sum_{i=1}^{k} \sum_{j=1}^{n} s_{ij} \beta_j \\
&= \sum_{j=1}^{n} c_{kj} \beta_j,
\end{aligned}$$

where $c_{kj} = \sum_{i=1}^{k} s_{ij} \leq 1$, and $\sum_{j=1}^{n} c_{kj} = k$, since $\sum_{j=1}^{n} c_{kj}$ is the sum of the entries in the first k rows of a doubly stochastic matrix. Using Lemmas 2.1 and 2.2, we have

$$\begin{aligned}\sum_{j=1}^{n} c_{kj}\beta_j &\leq \sum_{j=1}^{n} c_{kj}^{*}\beta_j^{*} \\ &\leq \sum_{j=1}^{k} \beta_j^{*}.\end{aligned}$$

Hence

$$\sum_{i=1}^{k} \alpha_i^{*} \leq \sum_{i=1}^{k} \beta_i^{*},$$

for $k = 1, 2, \ldots, n-1$. Moreover,

$$\begin{aligned}\sum_{i=1}^{n} \alpha_i &= \sum_{i=1}^{n} \sum_{j=1}^{n} s_{ij}\beta_j \\ &= \sum_{j=1}^{n} \beta_j \sum_{i=1}^{n} s_{ij} \\ &= \sum_{j=1}^{n} \beta_j.\end{aligned}$$

Thus $\alpha \prec \beta$.

Now suppose that $\alpha \prec \beta$. We show that there exists a doubly stochastic matrix S such that $\alpha = S\beta$. Clearly, it is sufficient to prove that $\alpha^{*} = S\beta^{*}$ for some S in Ω_n. For, if $\alpha^{*} = P\alpha$ and $\beta^{*} = Q\beta$, where P and Q are permutation matrices, then $\alpha = (P^{\mathrm{T}}SQ)\beta$, and $P^{\mathrm{T}}SQ \in \Omega_n$. We can assume therefore that $\alpha = \alpha^{*}$ and $\beta = \beta^{*}$. Suppose that $\alpha^{*} \neq \beta^{*}$. Call the number of nonzero coordinates in $\beta^{*} - \alpha^{*}$ the *discrepancy* between α and β, and denote it by $\delta(\alpha, \beta)$. Since $\alpha^{*} \neq \beta^{*}$, it is clear that $\delta(\alpha, \beta) \geq 2$. Since $\sum_{i=1}^{n}(\alpha_i - \beta_i) = 0$ and not all these differences can be zero, some of them must be positive and some negative. Let t be the least subscript such that $\alpha_t > \beta_t$, and let s be the greatest subscript, less than t, for which $\alpha_s < \beta_s$. Thus we have

$$\alpha_s < \beta_s, \quad \alpha_{s+1} = \beta_{s+1}, \quad \alpha_{s+2} = \beta_{s+2}, \ldots, \alpha_{t-1} = \beta_{t-1}, \quad \alpha_t > \beta_t. \tag{2}$$

Let S_1 be the elementary doubly stochastic $n \times n$ matrix with θ in positions (s, s) and (t, t), and $1 - \theta$ in positions (s, t) and (t, s). Then

$$\begin{aligned}(S_1\beta)_s &= \theta\beta_s + (1-\theta)\beta_t, \\ (S_1\beta)_t &= (1-\theta)\beta_s + \theta\beta_t,\end{aligned}$$

and

$$(S_1\beta)_i = \beta_i,$$

for all other i. We choose two values for θ:

$$\theta_1 = \frac{\alpha_s - \beta_t}{\beta_s - \beta_t}, \qquad \theta_2 = \frac{\beta_s - \alpha_t}{\beta_s - \beta_t}.$$

Since $\beta_s > \alpha_s \geq \alpha_t > \beta_t$, both these values lie in the interval $(0, 1)$. If $\theta = \theta_1$, then

$$(S_1\beta)_s = \alpha_s \quad \text{and} \quad (S_1\beta)_t = \beta_s - \alpha_s + \beta_t;$$

and if $\theta = \theta_2$, then

$$(S_1\beta)_s = \beta_s - \alpha_t + \beta_t \quad \text{and} \quad (S_1\beta)_t = \alpha_t.$$

Thus in either case the discrepancy between α and $S_1\beta$ is less than $\delta(\alpha, \beta)$, provided that $\alpha \prec S_1\beta$. This will be the case for θ_1 if

$$\beta_{t-1} \geq \beta_s - \alpha_s + \beta_t \geq \beta_{t+1}, \tag{3}$$

and for θ_2 if

$$\beta_{s-1} \geq \beta_s - \alpha_t + \beta_t \geq \beta_{s+1}. \tag{4}$$

Since $\beta_t + (\beta_s - \alpha_s) > \beta_t \geq \beta_{t+1}$ and $\beta_s - (\alpha_t - \beta_t) < \beta_s \leq \beta_{s-1}$, the right inequality in (3) and the left inequality in (4) clearly hold. Suppose that the left inequality in (3) fails to hold, that is,

$$\beta_{t-1} < \beta_s - \alpha_s + \beta_t.$$

Then

$$\begin{aligned}\beta_s - \alpha_t + \beta_t &> \beta_{t-1} + \alpha_s - \alpha_t \\ &= \alpha_{t-1} + \alpha_s - \alpha_t \\ &\geq \alpha_s \\ &\geq \alpha_{s+1} \\ &= \beta_{s+1},\end{aligned}$$

and (4) holds. Similarly, we can show that if the right inequality in (4) does not hold, then the left inequality in (3) must hold. We can conclude therefore that with the appropriate choice, $\theta = \theta_1$ or θ_2, we have

$$\delta(\alpha, S_1\beta) < \delta(\alpha, \beta), \qquad \alpha \prec S_1\beta, \quad \text{and} \quad (S_1\beta)^* = S_1\beta.$$

Continuing in the same fashion we can find a sequence of doubly stochastic matrices $S_1, S_2, \ldots, S_k$, such that the discrepancy $\delta(\alpha, S_kS_{k-1} \cdots S_1\beta)$ is zero, that is, $\alpha = S_k S_{k-1} \cdots S_1 \beta$. Now set $S = S_k S_{k-1} \cdots S_1$, which, by Lemma 1.1, is doubly stochastic. ■

The following example illustrates the method of the proof of Theorem 2.2: For given nonnegative 5-tuples α and β, satisfying $\alpha \prec \beta$, we find an "averaging" doubly stochastic matrix S such that $\alpha = S\beta$. The example also shows that substituting $S_1\beta$ for β may not reduce the discrepancy unless $(S_1\beta)^* = S_1\beta$.

Example 2.1. Let $\alpha = (9, 6, 5, 4, 4)$ and $\beta = (10, 10, 5, 2, 1)$. Then $\alpha \prec \beta$. Find a doubly stochastic matrix S such that $\alpha = S\beta$.

We use the notation in the proof of the preceding theorem. Here $t = 4$ and $s = 2$. Let

$$S_1 = \begin{bmatrix} 1 & 0 & 0 & 0 & 0 \\ 0 & \theta & 0 & 1-\theta & 0 \\ 0 & 0 & 1 & 0 & 0 \\ 0 & 1-\theta & 0 & \theta & 0 \\ 0 & 0 & 0 & 0 & 1 \end{bmatrix}.$$

If we take $\theta = \theta_1 = (\alpha_2 - \beta_4)/(\beta_2 - \beta_4) = \frac{1}{2}$, then $(S_1\beta)_2 = (S_1\beta)_4 = 6$, and $S_1\beta = (10, 6, 5, 6, 1)$. Thus $(S_1\beta)^* = (10, 6, 6, 5, 1)$, and $\delta(\alpha, S_1\beta) = \delta(\alpha, \beta) = 4$. We try $\theta = \theta_2 = (\beta_2 - \alpha_4)/(\beta_2 - \beta_4) = \frac{3}{4}$. Then $(S_1\beta)_4 = \alpha_4 = 4$, and $S_1\beta = (10, 8, 5, 4, 1)$. In this case, $\delta(\alpha, S_1\beta) = 3 < \delta(\alpha, \beta)$. Of course, this reduction in discrepancy was guaranteed by the theory.

We continue the process. Let

$$S_2 = \begin{bmatrix} 1 & 0 & 0 & 0 & 0 \\ 0 & \theta & 0 & 0 & 1-\theta \\ 0 & 0 & 1 & 0 & 0 \\ 0 & 0 & 0 & 1 & 0 \\ 0 & 1-\theta & 0 & 0 & \theta \end{bmatrix}.$$

Try $\theta = \theta_1' = (\alpha_2 - (S_1\beta)_5)/((S_1\beta)_2 - (S_1\beta)_5) = \frac{5}{7}$. Then $S_2S_1\beta = (10, 6, 5, 6, 1)$, and again we failed to progress: $\delta(\alpha, S_2S_1\beta) = \delta(\alpha, S_1\beta)$. The alternate choice, $\theta = \theta_2' = ((S_1\beta)_2 - \alpha_5)/((S_1\beta)_2 - (S_1\beta)_5) = \frac{4}{7}$, yields $S_2S_1\beta = (10, 5, 5, 4, 4)$, and with this value for θ we have $\delta(\alpha, S_2S_1\beta) = 2 < \delta(\alpha, S_1\beta)$.

Lastly, we set

$$S_3 = \begin{bmatrix} \theta & 1-\theta \\ 1-\theta & \theta \end{bmatrix} \dotplus I_3,$$

and $\theta = \theta_1'' = \frac{4}{5}$. Then $S_3S_2S_1\beta = (9,6,5,4,4) = \alpha$. Hence $\alpha = S\beta$, where

$$S = S_3S_2S_1 = \frac{1}{140}\begin{bmatrix} 112 & 12 & 0 & 4 & 12 \\ 28 & 48 & 0 & 16 & 48 \\ 0 & 0 & 140 & 0 & 0 \\ 0 & 35 & 0 & 105 & 0 \\ 0 & 45 & 0 & 15 & 80 \end{bmatrix}. \quad \blacksquare$$

We now proceed to prove Theorem 2.1 in the more general form of Hardy, Littlewood, and Pólya (see the remark preceding Theorem 2.2), in which α and β are assumed to be merely nonnegative n-tuples. We require the following lemma.

Lemma 2.3. *Let $c = (c_1, c_2, \ldots, c_n)$ be a positive n-tuple, and $\beta = (\beta_1, \beta_2, \ldots, \beta_n)$ be a nonnegative n-tuple. Let T be an elementary doubly stochastic matrix, and let $T\beta = \alpha = (\alpha_1, \alpha_2, \ldots, \alpha_n)$. Let $A(c)$ and $B(c)$ be as defined in the statement of Theorem 2.1. Then*

$$\operatorname{per}(A(c)) \le \operatorname{per}(B(c)).$$

Proof. We can assume without loss of generality that

$$T = \begin{bmatrix} \theta & 1-\theta \\ 1-\theta & \theta \end{bmatrix} \dotplus I_{n-2}.$$

Then

$$\begin{aligned}
&\operatorname{per}(B(c)) - \operatorname{per}(A(c)) \\
&\quad = \sum_{\sigma \in S_n} c_{\sigma(3)}^{\beta_3} c_{\sigma(4)}^{\beta_4} \cdots c_{\sigma(n)}^{\beta_n}\left(c_{\sigma(1)}^{\beta_1} c_{\sigma(2)}^{\beta_2} - c_{\sigma(1)}^{\theta\beta_1 + (1-\theta)\beta_2} c_{\sigma(2)}^{(1-\theta)\beta_1 + \theta\beta_2}\right) \\
&\quad = \sum_{\tau \in S_n} c_{\tau(3)}^{\beta_3} c_{\tau(4)}^{\beta_4} \cdots c_{\tau(n)}^{\beta_n}\left(c_{\tau(2)}^{\beta_1} c_{\tau(1)}^{\beta_2} - c_{\tau(2)}^{(1-\theta)\beta_1 + \theta\beta_2} c_{\tau(1)}^{\theta\beta_1 + (1-\theta)\beta_2}\right),
\end{aligned}$$

and therefore

$$\begin{aligned}
&2(\operatorname{per}(B(c)) - \operatorname{per}(A(c))) \\
&\quad = \sum_{\sigma \in S_n} c_{\sigma(3)}^{\beta_3} c_{\sigma(4)}^{\beta_4} \cdots c_{\sigma(n)}^{\beta_n}\Big(-c_{\sigma(1)}^{\theta\beta_1 + (1-\theta)\beta_2} c_{\sigma(2)}^{(1-\theta)\beta_1 + \theta\beta_2} \\
&\qquad\qquad - c_{\sigma(1)}^{(1-\theta)\beta_1 + \theta\beta_2} c_{\sigma(2)}^{\theta\beta_1 + (1-\theta)\beta_2} \\
&\qquad\qquad + c_{\sigma(1)}^{\beta_1} c_{\sigma(2)}^{\beta_2} + c_{\sigma(1)}^{\beta_2} c_{\sigma(2)}^{\beta_1}\Big) \\
&\quad = \sum_{\sigma \in S_n} c_{\sigma(1)}^{\beta_2} c_{\sigma(2)}^{\beta_2} c_{\sigma(3)}^{\beta_3} c_{\sigma(4)}^{\beta_4} \cdots c_{\sigma(n)}^{\beta_n}\left(c_{\sigma(1)}^{\theta(\beta_1 - \beta_2)} - c_{\sigma(2)}^{\theta(\beta_1 - \beta_2)}\right) \\
&\qquad \cdot \left(c_{\sigma(1)}^{(1-\theta)(\beta_1 - \beta_2)} - c_{\sigma(2)}^{(1-\theta)(\beta_1 - \beta_2)}\right) \ge 0. \quad \blacksquare
\end{aligned}$$

Corollary 2.1. *Let c, β, $A(c)$, and $B(c)$ be as defined in Lemma 2.3. If $S \in \Omega_n$ is a product of elementary matrices and $\alpha = S\beta$, then* $\operatorname{per}(A(c)) \leq \operatorname{per}(B(c))$.

Proof of Theorem 2.1. We assume that α and β are nonnegative n-tuples arranged in nonincreasing order. If $\alpha \prec \beta$, then, by Theorem 2.2, $\alpha = S\beta$, where S is a product of elementary doubly stochastic matrices (see the proof of Theorem 2.2), and thus

$$\operatorname{per}(A(c)) \leq \operatorname{per}(B(c)),$$

by Corollary 2.1.

To prove the converse assume that α and β are given, and

$$\operatorname{per}(A(c)) \leq \operatorname{per}(B(c)),$$

for all positive n-tuples $c = (c_1, c_2, \ldots, c_n)$. First, choose $c_1 = c_2 = \cdots = c_n = x > 0$. Then

$$\operatorname{per}(A(c)) = n!x^{\Sigma(\alpha, n)} \leq \operatorname{per}(B(c)) = n!x^{\Sigma(\beta, n)},$$

where $\Sigma(\gamma, m) = \Sigma_{i=1}^{m}\gamma_i$. Hence

$$x^{\Sigma(\alpha, n)} \leq x^{\Sigma(\beta, n)},$$

for all positive x (both greater than 1 and smaller than 1). Thus

$$\sum_{j=1}^{n} \alpha_j = \sum_{j=1}^{n} \beta_j.$$

Next, let $1 \leq k \leq n-1$, and set $c_1 = c_2 = \cdots = c_k = y > 1$, and $c_{k+1} = c_{k+2} = \cdots = c_n = 1$. Then

$$\operatorname{per}(A(c)) = ay^{\Sigma(\alpha, k)} + (\text{terms of degree lower than } \Sigma(\alpha, k)),$$

and

$$\operatorname{per}(B(c)) = by^{\Sigma(\beta, k)} + (\text{terms of degree lower than } \Sigma(\beta, k)),$$

where a and b are constants. But

$$\operatorname{per}(A(c)) \leq \operatorname{per}(B(c)),$$

for all y, and therefore for sufficiently large y we must have

$$y^{\Sigma(\alpha, k)} \leq y^{\Sigma(\beta, k)}.$$

It follows that

$$\Sigma(\alpha, k) \leq \Sigma(\beta, k),$$

that is,

$$\alpha_1 + \alpha_2 + \cdots + \alpha_k \leq \beta_1 + \beta_2 + \cdots + \beta_k,$$

for $k = 1, 2, \ldots, n-1$. Hence $\alpha \prec \beta$. ■

5.3. BIRKHOFF'S THEOREM

We now introduce one of the fundamental results in the theory of doubly stochastic matrices due to Birkhoff [2].

Theorem 3.1. *The set of $n \times n$ doubly stochastic matrices forms a convex polyhedron with permutation matrices as its vertices.*

In other words, if $A \in \Omega_n$, then

$$A = \sum_{j=1}^{s} \theta_j P_j, \tag{1}$$

where $P_1, P_2, \ldots, P_s$ are permutation matrices, and the θ_j are nonnegative numbers satisfying $\sum_{j=1}^{s} \theta_j = 1$.

Proof. Use induction on $\pi(A)$, the number of positive entries in A. If $\pi(A) = n$, then A is a permutation matrix, and the theorem holds with $s = 1$. Assume that $\pi(A) > n$ and that the theorem is true for all matrices in Ω_n with less than $\pi(A)$ positive entries. By Theorem 1.1, the matrix A has a positive diagonal $(a_{\sigma(1),1}, a_{\sigma(2),2}, \ldots, a_{\sigma(n),n})$, where $\sigma \in S_n$. Let $P = (p_{ij})$ be the incidence matrix of the permutation σ [that is, P is the permutation matrix with 1 in positions $(\sigma(i), i)$, $i = 1, 2, \ldots, n$], and let $a_{\sigma(t),t} = \min_i(a_{\sigma(i),i}) = a$. Clearly, $0 < a < 1$, since $a = 1$ would imply that A has 1 in positions $(\sigma(i), i)$, $i = 1, 2, \ldots, n$, and A would be a permutation matrix. Also, $A - aP$ is a nonnegative matrix because of the minimality of a. We assert that the matrix

$$B = (b_{ij}) = \frac{1}{1-a}(A - aP) \tag{2}$$

is doubly stochastic. Indeed,

$$\begin{aligned}
\sum_{j=1}^{n} b_{ij} &= \sum_{j=1}^{n} (a_{ij} - ap_{ij})/(1-a) \\
&= \left(\left(\sum_{j=1}^{n} a_{ij}\right) - a\left(\sum_{j=1}^{n} p_{ij}\right)\right)/(1-a) \\
&= (1-a)/(1-a) \\
&= 1, \qquad i = 1, 2, \ldots, n.
\end{aligned}$$

We can show similarly that

$$\sum_{i=1}^{n} b_{ij} = 1, \qquad j = 1, 2, \ldots, n.$$

Now, $\pi(B) \leq \pi(A) - 1$, since B has zero entries in all positions in which A has zeros, and, in addition, $b_{\sigma(t),t} = 0$. Hence, by the induction hypothesis,

$$B = \sum_{j=1}^{s-1} \gamma_j P_j,$$

where the P_j are permutation matrices, $\gamma_j \geq 0$, $j = 1, 2, \ldots, s-1$, and $\Sigma_{j=1}^{s-1}\gamma_j = 1$. But then, by (2),

$$\begin{aligned} A &= (1-a)B + aP \\ &= \left(\sum_{j=1}^{s-1} (1-a)\gamma_j P_j \right) + aP \\ &= \sum_{j=1}^{s} \theta_j P_j, \end{aligned}$$

where $\theta_j = (1-a)\gamma_j$ for $j = 1, 2, \ldots, s-1$, $\theta_s = a$, and $P_s = P$. Obviously, the θ_j are nonnegative. It remains to show that $\Sigma_{j=1}^{s}\theta_j = 1$. We compute

$$\begin{aligned} \sum_{j=1}^{s} \theta_j &= \left(\sum_{j=1}^{s-1} (1-a)\gamma_j \right) + a \\ &= (1-a)\left(\sum_{j=1}^{s-1} \gamma_j \right) + a \\ &= (1-a) + a \\ &= 1. \quad \blacksquare \end{aligned}$$

Let Λ_n^k denote the set of n-square (0,1)-matrices with k 1's in each row and each column. Such matrices occur in many combinatorial problems. If $A \in \Lambda_n^k$, then A/k is clearly doubly stochastic, and for this reason matrices in Λ_n^k are often called *doubly stochastic* (0,1)*-matrices.*

The following result, due to König [9], is an analogue of Theorem 3.1 for doubly stochastic (0, 1)-matrices. Historically, König's result preceded Birkhoff's theorem.

Theorem 3.2. *If* $A \in \Lambda_n^k$, *then*

$$A = \sum_{j=1}^{k} P_j, \tag{3}$$

where the P_j *are permutation matrices.*

The proof of Theorem 3.2 is quite straightforward; it follows the lines of the proof of Theorem 3.1. We leave it as an exercise (see Problem 10).

Birkhoff's theorem gives rise to the following tantalizing combinatorial problems:

(i) In how many ways can a given doubly stochastic matrix be expressed in the form (1)?

(ii) What is the least possible number of permutations in the representation of a doubly stochastic matrix A in the form (1)? In other words, what is the least number $\beta(A)$ of permutation matrices whose convex combination equals A?

Both problems are very hard. Practically nothing is known about question (i). Problem (ii) was proposed by Farahat and Mirsky [5], and some upper bounds for the number $\beta(A)$ have been obtained.

The first upper bound for $\beta(A)$ was given by Marcus and Newman [13] who deduced it from the procedure used in the proof of Theorem 3.1. This procedure consists of "stripping" multiples of permutation matrices, one by one, off a given $n \times n$ doubly stochastic matrix A, in such a way that at each stage at least one additional zero entry is produced. Thus, after no more than $n(n-1)$ such stages, the resulting doubly stochastic matrix has exactly n nonzero entries, and is therefore a permutation matrix. Hence A is a convex combination of at most $n(n-1)+1$ permutation matrices. It follows that

$$\beta(A) \leq n^2 - n + 1, \tag{4}$$

for any $A \in \Omega_n$. However, as we shall see, equality cannot hold in (4) for any $A \in \Omega_n$, $n > 1$. The following bound improves the bound in (4).

Theorem 3.3. *If $A \in \Omega_n$, then*

$$\beta(A) \leq (n-1)^2 + 1. \tag{5}$$

Proof. The dimension of the linear space of $n \times n$ real matrices is n^2. There are $2n$ linear conditions on the row sums and the column sums of $n \times n$ doubly stochastic matrices. Of these only $2n-1$ are independent, since the sum of all row sums in a matrix is necessarily equal to the sum of all its column sums. Hence

$$\begin{aligned}\dim \Omega_n &= n^2 - (2n-1)\\ &= (n-1)^2,\end{aligned}$$

and it follows, by Carathéodory's theorem (see, e.g., [20]), that every matrix A in Ω_n is in a convex hull of $(n-1)^2+1$ permutation matrices. Consequently, $\beta(A)$ cannot exceed $(n-1)^2+1$. ■

We now consider irreducible doubly stochastic matrices, and improve the upper bound in (5) by using the index of imprimitivity. We shall first require the following preliminary result.

Theorem 3.4 (Marcus, Minc, and Moyls [12]). *Let* $S = \dot{\Sigma}_{i=1}^{m} S_i$, *where* $S_i \in \Omega_{n_i}$, $i = 1, 2, \ldots, m$. *Then*

$$\beta(S) \leq \sum_{i=1}^{m} \beta(S_i) - m + 1. \tag{6}$$

Proof. Use induction on m. For $m = 2$ we have to show that

$$\beta(S_1 \dotplus S_2) \leq \beta(S_1) + \beta(S_2) - 1. \tag{7}$$

Let $S_1 = \Sigma_{i=1}^{r} \theta_i P_i$ and $S_2 = \Sigma_{j=1}^{s} \varphi_j Q_j$, where the P_i and the Q_j are $n_1 \times n_1$ and $n_2 \times n_2$ permutation matrices, respectively; $0 < \theta_1 \leq \theta_2 \leq \cdots \leq \theta_r$, $0 < \varphi_1 \leq \varphi_2 \leq \cdots \leq \varphi_s$; $\Sigma_{i=1}^{r} \theta_i = \Sigma_{j=1}^{s} \varphi_j = 1$; and $r = \beta(S_1)$, $s = \beta(S_2)$. We use induction on $r + s$. If $r + s = 2$, then $S_1 = P_1$, $S_2 = Q_1$, and $S_1 \dotplus S_2 = P_1 \dotplus Q_1$, which is a permutation matrix, and therefore (7) holds. Now, suppose that $r + s > 2$. We can assume, without loss of generality, that $\theta_1 \leq \varphi_1$. Then

$$S_1 \dotplus S_2 = \theta_1 (P_1 \dotplus Q_1) + (1 - \theta_1)\left(\left(\sum_{i=2}^{r} \frac{\theta_i}{1 - \theta_1} P_i\right) \dotplus \left(\frac{\varphi_1 - \theta_1}{1 - \theta_1} Q_1 + \sum_{j=2}^{s} \frac{\varphi_j}{1 - \theta_1} Q_j\right)\right).$$

Clearly,

$$\sum_{i=2}^{r} \frac{\theta_i}{1 - \theta_1} P_i \in \Omega_{n_1} \quad \text{and} \quad \frac{\varphi_1 - \theta_1}{1 - \theta_1} Q_1 + \sum_{j=2}^{s} \frac{\varphi_j}{1 - \theta_1} Q_j \in \Omega_{n_2}.$$

Thus

$$S_1 \dotplus S_2 = \theta_1 (P_1 \dotplus Q_1) + (1 - \theta_1) R,$$

where R is a direct sum of two doubly stochastic matrices, the first of which is a convex combination of $r - 1$ permutation matrices, and the second is a convex combination of s permutation matrices. Hence, by the induction hypothesis applied to R, $\beta(R) \leq r + s - 2$, and therefore

$$\beta(S_1 \dotplus S_2) \leq r + s - 1.$$

This proves the theorem for $m = 2$. Let $m > 2$, and assume that the theorem holds for direct sums of $m - 1$ matrices. Then

$$\begin{aligned}\beta(S) &= \beta\left(\sum_{i=1}^{m}{}^{\cdot}\, S_i\right)\\ &\le \beta\left(\sum_{i=1}^{m-1}{}^{\cdot}\, S_i\right) + \beta(S_m) - 1\\ &\le \sum_{i=1}^{m-1} \beta(S_i) - (m-1) + 1 + \beta(S_m) - 1\\ &= \sum_{i=1}^{m} \beta(S_i) - m + 1. \quad \blacksquare\end{aligned}$$

Theorem 3.5 (Marcus, Minc, and Moyls [12]). *If A is an irreducible doubly stochastic $n \times n$ matrix with index of imprimitivity h, then*

$$\beta(A) \le h\left(\frac{n}{h} - 1\right)^2 + 1. \tag{8}$$

Proof. By Theorem 1.4, the index h divides n. Let $n = qh$ and let R be the $n \times n$ permutation matrix with 1's in positions (i, j) for i, j satisfying $i - j \equiv q \mod n$. Let P be a permutation matrix such that PAP^{T} is in the Frobenius form (see Theorem 1.4) with q-square blocks $A_{12}, A_{23}, \ldots, A_{h-1,h}, A_{h1}$ in the superdiagonal. Then

$$PAP^{\mathrm{T}}R = A_{12} \dotplus A_{23} \dotplus \cdots \dotplus A_{h-1,h} \dotplus A_{h1},$$

and thus, by Theorem 3.4,

$$\begin{aligned}\beta(A) &= \beta(PAP^{\mathrm{T}}R)\\ &\le \beta(A_{12}) + \beta(A_{23}) + \cdots + \beta(A_{h-1,h}) + \beta(A_{h1}) - h + 1.\end{aligned}$$

But from Theorem 3.3 we have

$$\beta(A_{i,i+1}) \le (q-1)^2 + 1, \qquad i = 1, 2, \ldots, h-1,$$

and

$$\beta(A_{h1}) \le (q-1)^2 + 1.$$

Therefore

$$\begin{aligned}\beta(A) &\le h\left((q-1)^2 + 1\right) - h + 1\\ &= h\left(\frac{n}{h} - 1\right)^2 + 1. \quad \blacksquare\end{aligned}$$

The preceding result together with Corollary 1.6 may be used to yield a better estimate for $\beta(A)$ than that given by a direct application of formula (8). This is illustrated in the following example.

Example 3.1 (Marcus, Minc, and Moyls [12]). Let

$$A = \begin{bmatrix} 0_4 & J_4 \\ S & 0_4 \end{bmatrix} \in \Omega_8,$$

where

$$S = \begin{bmatrix} 0_2 & I_2 \\ J_2 & 0_2 \end{bmatrix},$$

and 0_t denotes the $t \times t$ zero matrix. Estimate the value of $\beta(A)$.

Formula (5) gives

$$\beta(A) \le (8-1)^2 + 1 = 50.$$

We observe, however, that A is in a superdiagonal block form and $J_4 S$ is positive, and therefore, by Corollary 4.1, Chapter III, the matrix A is irreducible with index of imprimitivity 2. Thus from (8) we have

$$\beta(A) \le 2\left(\frac{8}{2} - 1\right)^2 + 1 = 19.$$

Now, we permute the rows and columns of A so that it becomes $J_4 \dotplus S$. Further permutations of rows and columns yield the matrix $J_4 \dotplus I_2 \dotplus J_2 = J_4 \dotplus I_1 \dotplus I_1 \dotplus J_2$. Hence, by Theorem 3.4,

$$\beta(A) \le \beta(J_4) + \beta(I_1) + \beta(I_1) + \beta(J_2) - 4 + 1. \tag{9}$$

A direct application of Theorem 3.3 to J_4 gives $\beta(J_4) \le 10$, and thus (9) implies that

$$\beta(A) \le 11.$$

However, it is obvious by inspection that $\beta(J_4) = 4$, and therefore (9) gives

$$\beta(A) \le 5.$$

In fact, it is not hard to show that actually $\beta(A) = 4$. ■

5.4. MORE ABOUT DOUBLY STOCHASTIC MATRICES

The inverse of a nonnegative matrix is nonnegative if and only if the matrix is a generalized permutation matrix (Lemma 1.1, Chapter I). It follows that the inverse of a doubly stochastic matrix is doubly stochastic if and only if the matrix is a permutation matrix. An entirely different result holds for doubly quasi-stochastic matrices.

Lemma 4.1. *The inverse of a nonsingular doubly quasi-stochastic matrix is doubly quasi-stochastic.*

In particular, the inverse of a doubly stochastic matrix is doubly quasi-stochastic.

Proof. Let A be a nonsingular $n \times n$ doubly stochastic matrix. Then

$$J_n = J_n I_n = J_n A A^{-1} = J_n A^{-1},$$

since A is doubly stochastic, and

$$J_n = I_n J_n = A^{-1} A J_n = A^{-1} J_n.$$

Hence A^{-1} is doubly quasi-stochastic. ■

The following corollary is an immediate consequence of the lemma.

Corollary 4.1. *If A and X are $n \times n$ doubly stochastic matrices and X is nonsingular, then XAX^{-1} is doubly quasi-stochastic.*

Of course, the matrix XAX^{-1} in Corollary 4.1 need not be nonnegative (see Problem 15). Also, if A is doubly stochastic and X is a nonsingular matrix such that XAX^{-1} is doubly stochastic, then obviously X may not be doubly stochastic or even doubly quasi-stochastic. For example, if $A = I_n$, then X may be any nonsingular $n \times n$ matrix. However, if A happens to be irreducible, then the following rather unexpected result holds.

Theorem 4.1 (Marcus, Minc, and Moyls [12]). *If A is an irreducible doubly stochastic $n \times n$ matrix and $B = XAX^{-1}$ is doubly stochastic, then X is a multiple of a doubly quasi-stochastic matrix. Moreover, there exists a doubly stochastic matrix Y such that $YAY^{-1} = B$.*

Proof. Let J be the $n \times n$ matrix all of whose entries are 1, and let $r_i(M)$ denote the ith row sum of matrix M. Since $XA = BX$, we have $XAJ = BXJ$, and therefore $XJ = B(XJ)$. But each column of XJ is equal to the n-tuple

$u(X) = (r_1(X), r_2(X), \ldots, r_n(X))$, and thus

$$Bu(X) = u(X).$$

Now, 1 is a simple eigenvalue of A and of B, since A is an irreducible doubly stochastic matrix, and therefore $u(X)$ must be a scalar multiple of the n-tuple e all of whose entries are 1. Hence $r_i(X) = \alpha$, $i = 1, 2, \ldots, n$, and, by a similar argument, we can prove that $r_i(X^{\mathrm{T}}) = \beta$, $i = 1, 2, \ldots, n$. But then $\alpha = \beta$, $JX = XJ = \alpha J$, and X is a scalar multiple of a doubly quasi-stochastic matrix.

Vector e is an eigenvector of both X and J corresponding to eigenvalues α and n; respectively. Hence $\alpha + kn$ is an eigenvalue of $X + kJ$ for any scalar k. Choose k so that $X + kJ$ is positive and nonsingular, and $\alpha + kn > 0$. Let $Y = (X + kJ)/(\alpha + kn)$. Then Y is doubly stochastic, and

$$\begin{aligned} YAY^{-1} &= (X + kJ)A(X + kJ)^{-1} \\ &= (BX + kJ)(X + kJ)^{-1} \\ &= B(X + kJ)(X + kJ)^{-1} \\ &= B. \quad \blacksquare \end{aligned}$$

If A is a positive semidefinite matrix, then there exists a unique positive semidefinite matrix B such that $B^2 = A$. The matrix B is called the *square root* of A, and is denoted by $A^{1/2}$. The square root of a positive semidefinite doubly stochastic matrix is not, in general, doubly stochastic. For example, the square root of the matrix

$$\frac{1}{4}\begin{bmatrix} 3 & 0 & 1 \\ 0 & 3 & 1 \\ 1 & 1 & 2 \end{bmatrix}$$

is the doubly quasi-stochastic matrix

$$\frac{1}{12}\begin{bmatrix} 5 + 3\sqrt{3} & 5 - 3\sqrt{3} & 2 \\ 5 - 3\sqrt{3} & 5 + 3\sqrt{3} & 2 \\ 2 & 2 & 8 \end{bmatrix},$$

which is not doubly stochastic. The following theorem characterizes square roots of doubly stochastic matrices.

Theorem 4.2 (Marcus and Minc [11]). *The square root of a positive semidefinite doubly stochastic matrix $A = (a_{ij})$ is doubly quasi-stochastic. If $a_{ii} \le 1/(n - 1)$, $i = 1, 2, \ldots, n$, then $A^{1/2}$ is doubly stochastic.*

Proof. Let $1, \lambda_2, \lambda_3, \ldots, \lambda_n$ be the (nonnegative) eigenvalues of A, and let $U = (u_{ij})$ be an orthogonal matrix for which

$$U^{\mathrm{T}}AU = \operatorname{diag}(1, \lambda_2, \lambda_3, \ldots, \lambda_n),$$

and whose first column is $[1, 1, \ldots, 1]^{\mathrm{T}}/\sqrt{n}$. Let

$$B = (b_{ij}) = U \operatorname{diag}\left(1, \sqrt{\lambda_2}, \sqrt{\lambda_3}, \ldots, \sqrt{\lambda_n}\right) U^{\mathrm{T}}. \tag{1}$$

Then $B^2 = A$, and

$$\begin{aligned} \sum_{j=1}^{n} b_{ij} &= \sum_{j=1}^{n} \sum_{k=1}^{n} u_{ik} \sqrt{\lambda_k}\, u_{jk} \\ &= \sum_{k=1}^{n} u_{ik} \sqrt{\lambda_k} \sum_{j=1}^{n} u_{jk}. \end{aligned}$$

But U is an orthogonal matrix, and therefore for $k > 1$,

$$\begin{aligned} 0 &= \sum_{j=1}^{n} u_{j1} u_{jk} \\ &= \frac{1}{\sqrt{n}} \sum_{j=1}^{n} u_{jk}. \end{aligned}$$

Hence

$$\begin{aligned} \sum_{j=1}^{n} b_{ij} &= u_{i1} \sqrt{\lambda_1} \sum_{j=1}^{n} u_{j1} \\ &= \frac{1}{\sqrt{n}} \cdot 1 \cdot n \cdot \frac{1}{\sqrt{n}} \\ &= 1, \end{aligned}$$

for $i = 1, 2, \ldots, n$. Similarly, we can show

$$\sum_{i=1}^{n} b_{ij} = 1,$$

for $j = 1, 2, \ldots, n$. This proves the first part of the theorem.

Now, suppose that $a_{ii} \leq 1/(n-1)$, $i = 1, 2, \ldots, n$, and define B as in (1). We assert that B is now nonnegative and therefore it is doubly stochastic. For, if b_{pq} were negative for some p and q, then

$$\begin{aligned} a_{pp} &= \sum_{j=1}^{n} b_{pj}^2 \\ &> \sum_{j \neq q} b_{pj}^2 \\ &\geq \frac{1}{n-1} \left(\sum_{j \neq q} b_{pj} \right)^2 \\ &> \frac{1}{n-1}, \end{aligned} \tag{2}$$

since

$$\begin{aligned}\sum_{j \neq q} b_{pj} &= \sum_{j=1}^{n} b_{pj} - b_{pq} \\ &= 1 - b_{pq} \\ &> 1.\end{aligned}$$

However, (2) would contradict our assumption that $a_{ii} \leq 1/(n-1)$ for all i. ■

We conclude this section with a result of Marcus and Ree [14] which improves and generalizes Theorem 1.1. We require the following lemma.

Lemma 4.2. *Suppose that every entry of an $n \times n$ matrix either has a certain property P or does not have this property. Then a necessary and sufficient condition that at least k entries in each diagonal of the matrix have the property P is that the matrix contain an $s \times t$ submatrix composed entirely of entries with property P, where $s + t = n + k$.*

The lemma is just a restatement of Theorem 2.2, Chapter IV, where "property P" was specialized, essentially without loss of generality, to "is 0."

Theorem 4.3. *Let A be an $n \times n$ doubly stochastic matrix, and let m be an integer, $1 \leq m \leq n$. Then there exists a diagonal of A with at least m entries greater than or equal to*

$$\mu = \begin{cases} \dfrac{4k}{(n+k)^2}, & \text{if } m \text{ is odd}, \\[2ex] \dfrac{4k}{(n+k)^2 - 1}, & \text{if } m \text{ is even}, \end{cases}$$

where $k = n - m + 1$.

Proof. Suppose that every diagonal of A contains fewer than m entries greater than or equal to μ. That is, in every diagonal there are at least $n - m + 1 = k$ entries less than μ. Hence, by Lemma 4.2, the matrix A must contain an $s \times t$ submatrix M, where $s + t = n + k$ and every entry in M is less than μ. We can assume that A is of the form

$$s\left\{\left[\begin{array}{c|c} \overbrace{M}^{t} & B \\ \hline C & D \end{array}\right]\right..$$

Let $\sigma(X)$ denote the sum of all entries in matrix X. Then

$$\sigma(M) + \sigma(B) = s,$$
$$\sigma(M) + \sigma(C) = t,$$

and therefore

$$\begin{aligned} 2\sigma(M) + \sigma(B) + \sigma(C) &= s + t \\ &= n + k. \end{aligned}$$

Also,

$$n = \sigma(A) = \sigma(M) + \sigma(B) + \sigma(C) + \sigma(D).$$

Hence

$$\sigma(M) - \sigma(D) = k,$$

and therefore

$$\sigma(M) \geq k.$$

But we have assumed that $\sigma(M) < st\mu$, and therefore

$$\mu > \frac{\sigma(M)}{st} \geq \frac{k}{\max st},$$

where $\max st$ is the largest value st takes on subject to the condition $s + t = n + k$. Thus if m is odd, and therefore $n + k = 2n - m + 1$ is even, we have

$$\max st = (n + k)^2/4,$$

and if m is even, and therefore $n + k$ is odd, then

$$\max st = \big((n + k)^2 - 1\big)/4.$$

Hence if m is odd, then

$$\mu > \frac{4k}{(n + k)^2} = \frac{4k}{(2n - m + 1)^2}$$

and if m is even, then

$$\mu > \frac{4k}{(n + k)^2 - 1} = \frac{4k}{(2n - m + 1)^2 - 1},$$

which contradicts the definition of μ. ■

For the case $m = n$ we have the following result.

Corollary 4.2. *If $A = (a_{ij})$ is an $n \times n$ doubly stochastic matrix, then*

$$\max_{\tau \in S_n} \min_i a_{i,\tau i} \geq \begin{cases} \dfrac{4}{(n+1)^2}, & \text{if } n \text{ is odd,} \\ \dfrac{4}{n(n+2)}, & \text{if } n \text{ is even.} \end{cases}$$

Example 4.1. Show that the bound μ in Theorem 4.3 is the best possible by constructing, for each n and m, a matrix none of whose diagonals contains m entries greater than μ.

Let A be an $n \times n$ doubly stochastic matrix. If m is even, we partition A into four blocks,

$$A = \begin{bmatrix} A_{11} & A_{12} \\ A_{21} & A_{22} \end{bmatrix},$$

where A_{11} is $(n+k)/2 \times (n+k)/2$ and all its entries are equal to $\mu = 4k/(n+k)^2$, all the entries of A_{12} and of A_{21} are $2/(n+k)$, and A_{22} is zero. Then A contains a submatrix all of whose entries are equal to μ, and since $(n+k)/2 + (n+k)/2 = n+k$, we can conclude, using Lemma 4.2, that every diagonal of A has at least k entries equal to μ, and therefore no diagonal of A can have $n - k + 1 = m$ entries greater than μ.

If m is even and $\mu = 4k/((n+k)^2 - 1)$, we partition A in a similar way, where now A_{11} is $(n+k-1)/2 \times (n+k+1)/2$ and all its entries are equal to μ, all the entries of A_{12} are equal to $2/(n+k-1)$, those of A_{21} are equal to $2/(n+k+1)$, and A_{22} is again zero. The conclusion follows by the lemma in the same way as in the preceding case. ■

5.5. THE VAN DER WAERDEN CONJECTURE ≡ THE EGORYČEV–FALIKMAN THEOREM

In 1926 van der Waerden [24] posed the problem of determining the minimum of the permanent function in Ω_n, the polyhedron of doubly stochastic $n \times n$ matrices. In view of Corollary 1.1, this minimum is positive. It was conjectured that

$$\operatorname{per}(S) \geq n!/n^n, \tag{1}$$

for all $S \in \Omega_n$, and that equality holds in (1) if and only if $S = J_n$. The conjecture, known as the *van der Waerden conjecture*, remained unresolved for over half a century until Egoryčev [3] and Falikman [4] proved it independently. For the history of the conjecture and the details of the many partial solutions and related results, see [17], [18], [19] and [20].

In this section we prove inequality (1) together with the condition for equality. Our proof follows Egoryčev's proof with a few variations. Instead of using Alexandrov's inequality for mixed discriminants [1] from which Egoryčev deduced an inequality for permanents (Theorem 5.1), we shall obtain the latter directly from a lemma of Falikman.

Let $\mathbf{a}_1, \mathbf{a}_2, \ldots, \mathbf{a}_{n-2}$ be positive n-tuples. Define a bilinear form f by

$$f(\mathbf{x}, \mathbf{y}) = \operatorname{per}(\mathbf{a}_1, \mathbf{a}_2, \ldots, \mathbf{a}_{n-2}, \mathbf{x}, \mathbf{y}),$$

for all real n-tuples $\mathbf{x}$ and $\mathbf{y}$.

Lemma 5.1. *Let $A = (a_{ij})$ be a positive $n \times (n-1)$ matrix with columns $\mathbf{a}_1, \mathbf{a}_2, \ldots, \mathbf{a}_{n-1}$, and let $\mathbf{b} = (b_1, b_2, \ldots, b_n)$ be a real n-tuple. If $f(\mathbf{a}_{n-1}, \mathbf{b}) = 0$, then $f(\mathbf{b}, \mathbf{b}) \leq 0$. Moreover, $f(\mathbf{b}, \mathbf{b}) = 0$ if and only if $\mathbf{b} = 0$.*

Proof. Use induction on n. If $n = 2$, then $0 = f(\mathbf{a}_1, \mathbf{b}) = a_{11}b_2 + a_{21}b_1$ implies that $b_2 = -a_{21}b_1/a_{11}$, and therefore $f(\mathbf{b}, \mathbf{b}) = 2b_1b_2 = -2a_{21}b_1^2/a_{11} \leq 0$. Also, $f(\mathbf{b}, \mathbf{b}) = 0$ if and only if $b_1 = 0$, and thus if and only if $\mathbf{b} = 0$.

Assume now that $n > 2$ and that the result holds for $(n-1)$-tuples. Let $\mathbf{x} = (x_1, x_2, \ldots, x_n)$ be a real n-tuple. We show that $f(\mathbf{x}, \mathbf{e}_n) = 0$, where $\mathbf{e}_n = (0, 0, \ldots, 0, 1)$, implies that $f(\mathbf{x}, \mathbf{x}) < 0$, unless $\mathbf{x}$ is a multiple of $\mathbf{e}_n$ in which case, of course, $f(\mathbf{x}, \mathbf{x}) = 0$. Suppose that $\mathbf{x}$ is not a multiple of $\mathbf{e}_n$, and

$$f(\mathbf{x}, \mathbf{e}_n) = \operatorname{per}(\mathbf{a}'_1, \mathbf{a}'_2, \ldots, \mathbf{a}'_{n-2}, \mathbf{x}') = 0, \tag{2}$$

where $\mathbf{a}'_j = (a_{1j}, a_{2j}, \ldots, a_{n-1,j})$, $j = 1, 2, \ldots, n-2$, and $\mathbf{x}' = (x_1, x_2, \ldots, x_{n-1}) \neq 0$. Expanding $\operatorname{per}(\mathbf{a}_1, \mathbf{a}_2, \ldots, \mathbf{a}_{n-2}, \mathbf{x}, \mathbf{x})$ by the last row and using (2), we obtain

$$f(\mathbf{x}, \mathbf{x}) = \sum_{j=1}^{n-2} a_{nj} f_j(\mathbf{x}', \mathbf{x}'), \tag{3}$$

where $f_j(\mathbf{x}', \mathbf{x}') = \operatorname{per}(\mathbf{a}'_1, \ldots, \mathbf{a}'_{j-1}, \mathbf{a}'_{j+1}, \ldots, \mathbf{a}'_{n-2}, \mathbf{x}', \mathbf{x}')$. Now, by (2), $f_j(\mathbf{x}', \mathbf{a}'_j) = 0$, and clearly $f_j(\mathbf{a}'_j, \mathbf{a}'_j) > 0$, $j = 1, 2, \ldots, n-2$. Since it is assumed that $\mathbf{x}$ is not a multiple of $\mathbf{e}_n$, and therefore $\mathbf{x}' \neq 0$, it follows from the induction hypothesis that $f_j(\mathbf{x}', \mathbf{x}') < 0$, $j = 1, 2, \ldots, n-2$. Hence we can conclude from (3) that $f(\mathbf{x}, \mathbf{x}) < 0$. We have shown that if a vector $\mathbf{x}$ is not a multiple of $\mathbf{e}_n$, and $f(\mathbf{x}, \mathbf{e}_n) = 0$, then $f(\mathbf{x}, \mathbf{x}) < 0$. Since $f(\mathbf{a}_{n-1}, \mathbf{a}_{n-1}) > 0$, the preceding argument shows that $f(\mathbf{a}_{n-1}, \mathbf{e}_n)$ cannot vanish.

Let

$$\eta = -f(\mathbf{b}, \mathbf{e}_n)/f(\mathbf{a}_{n-1}, \mathbf{e}_n).$$

Then $f(\mathbf{b} + \eta \mathbf{a}_{n-1}, \mathbf{e}_n) = 0$, and therefore

$$f(\mathbf{b} + \eta \mathbf{a}_{n-1}, \mathbf{b} + \eta \mathbf{a}_{n-1}) \le 0,$$

that is,

$$f(\mathbf{b}, \mathbf{b}) + \eta^2 f(\mathbf{a}_{n-1}, \mathbf{a}_{n-1}) \le 0.$$

Hence $f(\mathbf{b}, \mathbf{b}) \le 0$. If $f(\mathbf{b}, \mathbf{b}) = 0$, then we must have $\eta = 0$. But if $\eta = 0$, then $f(\mathbf{b}, \mathbf{e}_n) = 0$, and therefore $\mathbf{b}$ is a multiple of $\mathbf{e}_n$, $\mathbf{b} = \tau \mathbf{e}_n$. Then

$$0 = f(\mathbf{a}_{n-1}, \mathbf{b}) = f(\mathbf{a}_{n-1}, \tau \mathbf{e}_n) = \tau f(\mathbf{a}_{n-1}, \mathbf{e}_n)$$

implies that $\tau = 0$, and thus $\mathbf{b} = 0$. ■

Our next theorem is a special case of Alexandrov's inequality for mixed discriminants. The following permanent version is due to Egoryčev.

Theorem 5.1. *Let $\mathbf{a}_1, \mathbf{a}_2, \ldots, \mathbf{a}_{n-1}$ be positive n-tuples, and let $\mathbf{a}_n$ be a real n-tuple. Then*

$$\begin{aligned}&(\operatorname{per}(\mathbf{a}_1, \ldots, \mathbf{a}_{n-1}, \mathbf{a}_n))^2 \\ &\qquad \ge \operatorname{per}(\mathbf{a}_1, \ldots, \mathbf{a}_{n-2}, \mathbf{a}_{n-1}, \mathbf{a}_{n-1}) \operatorname{per}(\mathbf{a}_1, \ldots, \mathbf{a}_{n-2}, \mathbf{a}_n, \mathbf{a}_n). \qquad (4)\end{aligned}$$

Equality can hold in (4) *if and only if $\mathbf{a}_{n-1}$ and $\mathbf{a}_n$ are linearly dependent.*

Proof. Let f denote the bilinear form in Lemma 5.1, and let $t = f(\mathbf{a}_{n-1}, \mathbf{a}_n)/f(\mathbf{a}_{n-1}, \mathbf{a}_{n-1})$. If $\mathbf{b} = \mathbf{a}_n - t\mathbf{a}_{n-1}$, then

$$f(\mathbf{a}_{n-1}, \mathbf{b}) = f(\mathbf{a}_{n-1}, \mathbf{a}_n) - tf(\mathbf{a}_{n-1}, \mathbf{a}_{n-1}) = 0.$$

Hence, by Lemma 5.1,

$$\begin{aligned}0 \ge f(\mathbf{b}, \mathbf{b}) &= f(\mathbf{b}, \mathbf{a}_n) - tf(\mathbf{b}, \mathbf{a}_{n-1}) \\ &= f(\mathbf{b}, \mathbf{a}_n) \\ &= f(\mathbf{a}_n, \mathbf{a}_n) - tf(\mathbf{a}_{n-1}, \mathbf{a}_n) \\ &= f(\mathbf{a}_n, \mathbf{a}_n) - (f(\mathbf{a}_{n-1}, \mathbf{a}_n))^2 / f(\mathbf{a}_{n-1}, \mathbf{a}_{n-1}). \qquad (5)\end{aligned}$$

Thus

$$(f(\mathbf{a}_{n-1}, \mathbf{a}_n))^2 \ge f(\mathbf{a}_{n-1}, \mathbf{a}_{n-1}) f(\mathbf{a}_n, \mathbf{a}_n),$$

which is equivalent to inequality (4).

By Lemma 5.1, equality can hold in (5) if and only if $\mathbf{b} = 0$, that is, if and only if $\mathbf{a}_n = t\mathbf{a}_{n-1}$. ■

It can be easily shown by a continuity argument, that inequality (4) also holds if $\mathbf{a}_1, \mathbf{a}_2, \ldots, \mathbf{a}_{n-1}$ are merely nonnegative n-tuples. Of course, the condition for equality is no longer valid in this case. Furthermore, it is clear that any two n-tuples $\mathbf{a}_q$ and $\mathbf{a}_t$, not just the last two, can be the designated n-tuples in Theorem 5.1, provided that one of them and all the remaining n-tuples are nonnegative. Therefore we can deduce the following corollary to Theorem 5.1.

Corollary 5.1. *If $A = (a_{ij})$ is a nonnegative $n \times n$ matrix, then*

$$(\text{per}(A))^2 \geq \left(\sum_{i=1}^{n} a_{iq} \text{per}(A(i|t)) \right) \left(\sum_{i=1}^{n} a_{it} \text{per}(A(i|q)) \right), \tag{6}$$

for any q and t, $1 \leq q < t \leq n$. If all the columns of A, except possibly column t, are positive, then equality can hold in (6) *if and only if column t is a multiple of column q.*

Before we introduce and prove our next theorems we shall prove several lemmas. An $n \times n$ doubly stochastic matrix A is called *minimizing* in Ω_n if

$$\text{per}(A) = \min\{\text{per}(S) | S \in \Omega_n\}.$$

Lemma 5.2 (Marcus and Newman [13]). *A minimizing matrix is fully indecomposable.*

Proof. Let A be a minimizing matrix in Ω_n. Suppose that A is partly decomposable. Then, by Corollary 1.4, there exist permutation matrices P and Q such that $PAQ = B \dotplus C$, where $B = (b_{ij}) \in \Omega_k$ and $C = (c_{ij}) \in \Omega_{n-k}$. We show that there exists an $n \times n$ doubly stochastic matrix whose permanent is less than per(A). Since, by Corollary 1.1, the permanent of A is positive, we can assume without loss of generality that $b_{kk}\text{per}((PAQ)(k|k)) > 0$ and $c_{11}\text{per}((PAQ)(k+1|k+1)) > 0$. Let ε be any positive number less than $\min\{b_{kk}, c_{11}\}$, and let

$$G(\varepsilon) = PAQ - \varepsilon(E_{kk} + E_{k+1,k+1}) + \varepsilon(E_{k,k+1} + E_{k+1,k}).$$

Then $G(\varepsilon) \in \Omega_n$, and

$$\begin{aligned}
\text{per}(G(\varepsilon)) &= \text{per}(PAQ) - \varepsilon\,\text{per}((PAQ)(k|k)) + \varepsilon\,\text{per}((PAQ)(k|k+1)) \\
&\quad - \varepsilon\,\text{per}((PAQ)(k+1|k+1)) + \varepsilon\,\text{per}((PAQ)(k+1|k)) + O(\varepsilon^2) \\
&= \text{per}(A) - \varepsilon(\text{per}((PAQ)(k|k)) \\
&\qquad + \text{per}((PAQ)(k+1|k+1))) + O(\varepsilon^2),
\end{aligned}$$

since $\text{per}((PAQ)(k|k+1)) = \text{per}((PAQ)(k+1|k)) = 0$, by the Frobenius–König theorem. Also, $\text{per}((PAQ)(k|k)) + \text{per}((PAQ)(k+1|k+1)) > 0$, and therefore for sufficiently small positive ε,

$$\text{per}(G(\varepsilon)) < \text{per}(A),$$

contradicting the assumption that A is a minimizing matrix. ■

Lemma 5.3 (Marcus and Newman [13]). *If $A = (a_{ij})$ is a minimizing matrix in Ω_n, then $a_{hk} > 0$ implies that* $\text{per}(A(h|k)) = \text{per}(A)$.

Proof. Let $C(A)$ be the face of Ω_n of least dimension containing A in its interior. In other words,

$$C(A) = \{X = (x_{ij}) \in \Omega_n | x_{ij} = 0 \text{ if } (i, j) \in Z\},$$

where $Z = \{(i, j) | a_{ij} = 0\}$. Then $C(A)$ is defined by the following conditions:

$$\sum_{j=1}^{n} x_{ij} = 1, \qquad i = 1, 2, \ldots, n,$$

$$\sum_{i=1}^{n} x_{ij} = 1, \qquad j = 1, 2, \ldots, n,$$

$$x_{ij} \geq 0, \qquad i, j = 1, 2, \ldots, n,$$

$$x_{ij} = 0, \qquad (i, j) \in Z.$$

Since A is in the interior of $C(A)$, and the permanent function has an absolute minimum at A, it must have a stationary point there. Hence we may introduce Lagrange multipliers and set up the function

$$F(X) = \text{per}(X) - \sum_{i=1}^{n} \lambda_i \left(\sum_{k=1}^{n} x_{ik} - 1 \right) - \sum_{j=1}^{n} \mu_j \left(\sum_{k=1}^{n} x_{kj} - 1 \right),$$

for $X \in C(A)$. Now, for $(i, j) \notin Z$,

$$\partial F(X) / \partial x_{ij} = \text{per}(X(i|j)) - \lambda_i - \mu_j.$$

Therefore

$$\text{per}(A(i|j)) = \lambda_i + \mu_j, \tag{7}$$

and thus

$$\begin{aligned}
\operatorname{per}(A) &= \sum_{j=1}^{n} a_{ij}\operatorname{per}(A(i|j)) \\
&= \sum_{j=1}^{n} a_{ij}(\lambda_i + \mu_j) \\
&= \lambda_i + \sum_{j=1}^{n} a_{ij}\mu_j, \qquad i = 1,2,\ldots,n. \qquad (8)
\end{aligned}$$

Similarly,

$$\begin{aligned}
\operatorname{per}(A) &= \sum_{i=1}^{n} a_{ij}\operatorname{per}(A(i|j)) \\
&= \mu_j + \sum_{i=1}^{n} a_{ij}\lambda_i, \qquad j = 1,2,\ldots,n. \qquad (9)
\end{aligned}$$

Now, let $\mathbf{e} = (1,1,\ldots,1)$, $\boldsymbol{\lambda} = (\lambda_1, \lambda_2, \ldots, \lambda_n)$, $\boldsymbol{\mu} = (\mu_1, \mu_2, \ldots, \mu_n)$. Then from (8) and (9) we have

$$\operatorname{per}(A)\,\mathbf{e} = \boldsymbol{\lambda} + A\boldsymbol{\mu}, \qquad (10)$$

$$\operatorname{per}(A)\,\mathbf{e} = A^{\mathrm{T}}\boldsymbol{\lambda} + \boldsymbol{\mu}. \qquad (11)$$

Premultiply (10) by A^{T}:

$$\operatorname{per}(A)\,\mathbf{e} = A^{\mathrm{T}}\boldsymbol{\lambda} + A^{\mathrm{T}}A\boldsymbol{\mu}, \qquad (12)$$

since $A^{\mathrm{T}}\mathbf{e} = \mathbf{e}$. Subtract (11) from (12):

$$A^{\mathrm{T}}A\boldsymbol{\mu} = \boldsymbol{\mu}.$$

Similarly,

$$AA^{\mathrm{T}}\boldsymbol{\lambda} = \boldsymbol{\lambda}.$$

Now, by Lemma 5.1, A is fully indecomposable, and therefore both $A^{\mathrm{T}}A$ and AA^{T} are fully indecomposable (see Problem 2, Chapter IV), and therefore each of them has 1 as a simple eigenvalue. Thus both $\boldsymbol{\lambda}$ and $\boldsymbol{\mu}$ are multiples of $\mathbf{e}$: $\boldsymbol{\lambda} = c\mathbf{e}$ and $\boldsymbol{\mu} = d\mathbf{e}$, say. It follows from (7) that

$$\operatorname{per}(A(i|j)) = c + d,$$

for all $(i, j) \notin Z$. Hence

$$\begin{aligned} \operatorname{per}(A) &= \sum_{j=1}^{n} a_{ij} \operatorname{per}(A(i|j)) \\ &= \sum_{j=1}^{n} a_{ij}(c + d) \\ &= c + d \\ &= \operatorname{per}(A(i|j)), \end{aligned}$$

for all $(i, j) \notin Z$. ■

Lemma 5.4 (London [10]). *If A is a minimizing matrix in Ω_n, then*

$$\operatorname{per}(A(i|j)) \geq \operatorname{per}(A),$$

for all i and j.

Proof (Minc [16]). Let $P = (p_{ij})$ be an $n \times n$ permutation matrix. For $0 \leq \theta \leq 1$ define the function

$$f_P(\theta) = \operatorname{per}((1 - \theta)A + \theta P).$$

Since A is a minimizing matrix,

$$f_P'(0) \geq 0,$$

for any permutation matrix P. But

$$\begin{aligned} f_P'(0) &= \sum_{s,t=1}^{n} (-a_{st} + p_{st}) \operatorname{per}(A(s|t)) \\ &= \sum_{s,t=1}^{n} p_{st} \operatorname{per}(A(s|t)) - n \operatorname{per}(A) \\ &= \sum_{s=1}^{n} \operatorname{per}(A(s|\sigma(s))) - n \operatorname{per}(A), \end{aligned}$$

where σ is the permutation corresponding to P. Hence

$$\sum_{s=1}^{n} \operatorname{per}(A(s|\sigma(s))) \geq n \operatorname{per}(A), \tag{13}$$

for any permutation σ. Now, by Lemma 5.2, the matrix A is fully indecomposable and therefore, by Theorem 4.1, Chapter IV, every entry of A lies on a diagonal all of whose other $n-1$ entries are positive. In other words, for any (i, j) there exists a permutation σ such that $j = \sigma(i)$ and $a_{s,\sigma(s)} > 0$ for $s = 1, \dots, i-1, i+1, \dots, n$. But this implies, by Lemma 5.3, that

$$\operatorname{per}(A(s|\sigma(s))) = \operatorname{per}(A), \tag{14}$$

for $s = 1, \dots, i-1, i+1, \dots, n$. Since $j = \sigma(i)$, it follows from (13) and (14) that

$$\operatorname{per}(A(i|j)) \geq \operatorname{per}(A). \quad \blacksquare$$

Lemma 5.5. *Let $A = (a_{ij})$ be an $n \times n$ matrix, and suppose that the corresponding permanental cofactors of entries in columns s and t, $s < t$, are equal:* $\operatorname{per}(A(i|s)) = \operatorname{per}(A(i|t))$, $i = 1, 2, \dots, n$. *Then the permanent of the matrix obtained from A by replacing column s and column t of A by their arithmetical mean is equal to the permanent of A:*

$$\begin{aligned}\operatorname{per}(\mathbf{a}_1, \dots, \mathbf{a}_{s-1}, (\mathbf{a}_s + \mathbf{a}_t)/2, \mathbf{a}_{s+1}, \dots, \mathbf{a}_{t-1}, (\mathbf{a}_s + \mathbf{a}_t)/2, \mathbf{a}_{t+1}, \dots, \mathbf{a}_n)\\ = \operatorname{per}(A).\end{aligned}$$

Proof. The result in this lemma is nearly obvious. For, by multilinearity of the permanent function, we have

$$\begin{aligned}
&\phantom{per(\mathbf{a}_1, \dots,} s \phantom{(\mathbf{a}_s + \mathbf{a}_t)/2, \dots, (\mathbf{a}_s + } t\\
&\operatorname{per}(\mathbf{a}_1, \dots, (\mathbf{a}_s + \mathbf{a}_t)/2, \dots, (\mathbf{a}_s + \mathbf{a}_t)/2, \dots, \mathbf{a}_n)\\
&= \operatorname{per}(\mathbf{a}_1, \dots, \mathbf{a}_s, \dots, \qquad \mathbf{a}_s, \dots, \mathbf{a}_n)/4\\
&+ \operatorname{per}(\mathbf{a}_1, \dots, \mathbf{a}_s, \dots, \qquad \mathbf{a}_t, \dots, \mathbf{a}_n)/4\\
&+ \operatorname{per}(\mathbf{a}_1, \dots, \mathbf{a}_t, \dots, \qquad \mathbf{a}_s, \dots, \mathbf{a}_n)/4\\
&+ \operatorname{per}(\mathbf{a}_1, \dots, \mathbf{a}_t, \dots, \qquad \mathbf{a}_t, \dots, \mathbf{a}_n)/4\\
&= \left(\sum_{i=1}^{n} a_{is} \operatorname{per}(A(i|t)) + \operatorname{per}(A) + \operatorname{per}(A) + \sum_{i=1}^{n} a_{it} \operatorname{per}(A(i|s)) \right) \Big/ 4\\
&= \left(\sum_{i=1}^{n} a_{is} \operatorname{per}(A(i|s)) + \operatorname{per}(A) + \operatorname{per}(A) + \sum_{i=1}^{n} a_{it} \operatorname{per}(A(i|t)) \right) \Big/ 4\\
&= \operatorname{per}(A). \quad \blacksquare
\end{aligned}$$

Marcus and Newman [13] hoped to prove that all permanental cofactors of a minimizing matrix are equal to the permanent of the matrix. This together with the "averaging process" in Lemma 5.5 would have proved the van der Waerden conjecture (cf. the proof in [13] of the theorem that a positive minimizing matrix must be equal to J_n). Egoryčev [3] used London's result (Lemma 5.4) together with his own theorem (Theorem 5.1) to obtain the key result on permanental cofactors of a minimizing doubly stochastic matrix. Our next theorem gives Egoryčev's result in a somewhat more general form.

A square nonnegative matrix is called *column (row) stochastic* if all its row (column) sums are 1.

Theorem 5.2. *Let A be an $n \times n$ column (row) stochastic matrix satisfying*

$$0 < \operatorname{per}(A) \leq \operatorname{per}(A(i|j)), \qquad i, j = 1, 2, \ldots, n. \tag{15}$$

Then

$$\operatorname{per}(A(i|j)) = \operatorname{per}(A), \qquad i, j = 1, 2, \ldots, n.$$

Proof. Let $A = (a_{ij})$ be a column stochastic matrix satisfying condition (15), and suppose that for some s and t the inequality in (15) is strict, that is,

$$\operatorname{per}(A(s|t)) > \operatorname{per}(A).$$

Let a_{sq} be a positive entry in the sth row of A, where $q \neq t$. Such an entry must exist, since the condition that $\operatorname{per}(A(i|j)) > 0$ for all i and j guarantees that A has at least two positive entries in each row. Then

$$a_{it}\operatorname{per}(A(i|q)) \geq a_{it}\operatorname{per}(A), \qquad i = 1, 2, \ldots, n,$$

and

$$a_{iq}\operatorname{per}(A(i|t)) \geq a_{iq}\operatorname{per}(A), \qquad i = 1, 2, \ldots, n.$$

If we had strict inequality for $i = s$,

$$a_{sq}\operatorname{per}(A(s|t)) > a_{sq}\operatorname{per}(A),$$

then it would follow, by Corollary 5.1, that

$$\begin{aligned}(\operatorname{per}(A))^2 &\geq \left(\sum_{i=1}^{n} a_{iq}\operatorname{per}(A(i|t))\right)\left(\sum_{i=1}^{n} a_{it}\operatorname{per}(A(i|q))\right) \\ &> \left(\sum_{i=1}^{n} a_{iq}\operatorname{per}(A)\right)\left(\sum_{i=1}^{n} a_{it}\operatorname{per}(A)\right) \\ &= (\operatorname{per}(A))^2.\end{aligned}$$

This contradiction proves that $\operatorname{per}(A(s|t))$ cannot be greater than $\operatorname{per}(A)$ for any s and t.

The proof for the case when A is row stochastic is similar. ■

Theorem 5.2, Corollary 1.1, and Lemma 5.4 yield immediately the following results.

Theorem 5.3 (Egoryčev [3]). *If A is a minimizing matrix in Ω_n, then*

$$\operatorname{per}(A(i|j)) = \operatorname{per}(A), \qquad i, j = 1, 2, \ldots, n.$$

Corollary 5.2. *If A is a minimizing matrix in Ω_n, then*

$$(\operatorname{per}(A))^2 = \left(\sum_{i=1}^{n} a_{iq}\operatorname{per}(A(i|t))\right)\left(\sum_{i=1}^{n} a_{it}\operatorname{per}(A(i|q))\right),$$

for any q and t, $1 \le q < t \le n$.

Corollary 5.3. *If A is a minimizing matrix in Ω_n, and B is the matrix obtained from A by replacing each of the two arbitrary columns of A by their arithmetical mean, then* $\operatorname{per}(B) = \operatorname{per}(A)$.

We are now ready to prove the van der Waerden conjecture.

Theorem 5.4. *If S is a doubly stochastic $n \times n$ matrix and $S \neq J_n$, then*

$$\operatorname{per}(S) > \operatorname{per}(J_n) = n!/n^n.$$

Proof. Let A be a minimizing matrix in Ω_n. We show that $A = J_n$. By Lemma 5.2, the matrix A is fully indecomposable, and therefore it has at least two positive entries in each of its rows. Consider the jth column of A. Applying the averaging process of Lemma 5.5 to pairs of columns of A, other than the jth column, we can obtain, after a finite number of steps, a doubly stochastic matrix C all of whose columns, except possibly the jth column, are positive. By Corollary 5.3, $\operatorname{per}(C) = \operatorname{per}(A)$. Thus C is also a minimizing matrix in Ω_n, and therefore for any integer i, $1 \le i \le n$, $i \neq j$, we have, by Corollary 5.3,

$$(\operatorname{per}(C))^2 = \left(\sum_{t=1}^{n} c_{ti}\operatorname{per}(C(t|j))\right)\left(\sum_{t=1}^{n} c_{tj}\operatorname{per}(C(t|i))\right).$$

It follows, by virtue of Corollary 5.1, that the ith and the jth columns of C are equal. But this is true for every i, $i \neq j$. Hence all the columns of C are equal, and therefore $C = J_n$. Thus the jth column of C, which is the jth column of A, is a column of J_n. Since this is true for any j, the matrix A must be equal to J_n. ■

PROBLEMS

1 Let $A = (a_{ij})$ be an $n \times n$ (0,1)-matrix with a positive permanent. Define the $n \times n$ matrix $B = (b_{ij})$ by

$$b_{ij} = a_{ij}\operatorname{per}(A(i|j))/\operatorname{per}(A), \qquad i, j = 1, 2, \ldots, n.$$

Show that B is doubly stochastic.

2 Show that every doubly stochastic 2×2 matrix is orthostochastic.

3 Which of the following matrices are not Schur-stochastic? Orthostochastic?

$$\frac{1}{16}\begin{bmatrix} 8 & 7 & 1 \\ 7 & 2 & 7 \\ 1 & 7 & 8 \end{bmatrix}, \quad \frac{1}{12}\begin{bmatrix} 4 & 4 & 4 \\ 3 & 6 & 3 \\ 5 & 2 & 5 \end{bmatrix}, \quad \frac{1}{3}\begin{bmatrix} 0 & 2 & 2 \\ 2 & 1 & 1 \\ 2 & 1 & 1 \end{bmatrix}, \quad \frac{1}{3}\begin{bmatrix} 0 & 1 & 1 & 1 \\ 1 & 0 & 1 & 1 \\ 1 & 1 & 0 & 1 \\ 1 & 1 & 1 & 0 \end{bmatrix}.$$

4 Show that 1 is the maximal eigenvalue of any row stochastic matrix, and that $\mathbf{e} = (1, 1, \ldots, 1)$ is a corresponding eigenvector. Are all the eigenvectors corresponding to 1 multiples of $\mathbf{e}$?

5 Prove Corollary 1.3.

6 Prove Corollaries 1.4, 1.5, and 1.6.

7 Show that the matrix

$$\frac{1}{2}\begin{bmatrix} 1 & 1 & 0 \\ 0 & 1 & 1 \\ 1 & 0 & 1 \end{bmatrix}$$

is not a product of elementary doubly stochastic matrices.

8 Let $\alpha = (12, 13, 11)$ and $\beta = (6, 18, 12)$.

(a) Let $A(c)$ and $B(c)$ be the matrices defined in Theorem 2.1, where $c = (3, 2, 1)$, and α and β are the triples defined above. Verify that $\operatorname{per}(A(c)) \leq \operatorname{per}(B(c))$.

(b) Find a doubly stochastic matrix S such that $\alpha = S\beta$.

9 Let $\alpha = (9, 7, 5, 4, 4)$ and $\beta = (10, 10, 5, 3, 1)$. Find a doubly stochastic matrix S such that $\alpha = S\beta$.

10 Prove Theorem 3.2.

11 Let

$$A = \begin{bmatrix} 0 & C \\ I_4 & 0 \end{bmatrix} \in \Omega_8 \quad \text{and} \quad B = \begin{bmatrix} 0 & C \\ D & 0 \end{bmatrix} \in \Omega_8,$$

where

$$C = \frac{1}{2}\begin{bmatrix} 0 & 1 & 0 & 1 \\ 1 & 0 & 1 & 0 \\ 0 & 1 & 0 & 1 \\ 1 & 0 & 1 & 0 \end{bmatrix} \quad \text{and} \quad D = \begin{bmatrix} 0 & 1 & 0 & 0 \\ 1 & 0 & 0 & 0 \\ 0 & 0 & 0 & 1 \\ 0 & 0 & 1 & 0 \end{bmatrix}.$$

(a) Determine the indices of imprimitivity of A and B.

(b) Use Theorem 3.5 to obtain upper bounds for $\beta(A)$ and $\beta(B)$.

(c) Express C and D as convex combinations of permutation matrices, and determine the actual values of $\beta(A)$ and $\beta(B)$.

12 Show without the use of Theorem 4.3 that every 3×3 doubly stochastic matrix contains a diagonal all of whose entries are greater than or equal to $\frac{1}{4}$. For any $k > \frac{1}{4}$ construct a 3×3 doubly stochastic matrix each diagonal of which contains an entry less than k.

13 Prove without the use of Theorem 4.3 that every 4×4 doubly stochastic matrix contains a diagonal all of whose entries are at least $\frac{1}{6}$.

14 Let $A = (a_{ij})$ be a positive semidefinite doubly stochastic $n \times n$ matrix. Show that the condition in Theorem 4.2,

$$a_{ii} \le 1/(n-1), \qquad i = 1, 2, \ldots, n,$$

is not necessary for A to have a doubly stochastic square root, even if A happens to be positive.

15 Construct doubly stochastic matrices X and A such that XAX^{-1} is not doubly stochastic.

16 Find a minimizing matrix in the set of all doubly stochastic 3×3 matrices with zero trace.

17 Find a minimizing matrix in the set of all doubly stochastic 3×3 matrices with 0's in positions (1, 1) and (2, 2).

REFERENCES

1. A. D. Alexandrov, On the theory of mixed volumes of convex bodies IV, *Mat. Sb. (N.S.)* **3**(45) (1938), 227–251 (in Russian).

2. G. Birkhoff, Tres observaciones sobre el algebra lineal, *Univ. Nac. Tucumán Rev. Ser. A* **5** (1946), 147–150.

3. G. P. Egoryčev, A solution of van der Waerden's permanent problem, *Dokl. Akad. Nauk SSSR* **258** (1981), 1041–1044 (in Russian). Translated in *Soviet Math. Dokl.* **23** (1981), 619–622.

4. D. I. Falikman, A proof of van der Waerden's conjecture on the permanent of a doubly stochastic matrix, *Mat. Zametki* **29** (1981), 931–938 (in Russian). Translated in *Math. Notes* **29** (1981), 475–479.
5. H. K. Farahat and L. Mirsky, Permutation endomorphisms and refinement of a theorem of Birkhoff, *Proc. Cambridge Philos. Soc.* **56** (1960), 322–328.
6. G. H. Hardy, J. E. Littlewood, and G. Pólya, *Inequalities*, Cambridge University Press, London, 1934.
7. D. Knuth, A permanental inequality, *Amer. Math. Monthly* **88** (1981), 731–740.
8. D. König, Über Graphen und ihre Anwendung auf Determinantentheorie und Mengenlehre, *Math. Ann.* **77** (1916), 453–465.
9. D. König, *Theorie der endlichen und unendlichen Graphen*, Akademische Verlagsgesellschaft, Leipzig, 1936.
10. D. London, Some notes on the van der Waerden conjecture, *Linear Algebra Appl.* **4** (1971), 155–160.
11. M. Marcus and H. Minc, Some results on doubly stochastic matrices, *Proc. Amer. Math. Soc.* **76** (1962), 571–579.
12. M. Marcus, H. Minc, and B. Moyls, Some results on nonnegative matrices, *J. Res. Nat. Bur. Standards Sect. B* **65** (1961), 205–209.
13. M. Marcus and M. Newman, On the minimum of the permanent of a doubly stochastic matrix, *Duke Math. J.* **26** (1959), 61–72.
14. M. Marcus and R. Ree, Diagonals of doubly stochastic matrices, *Quart. J. Math. Oxford Ser. (2)* **10** (1959), 295–302.
15. A. W. Marshall and I. Olkin, *Inequalities: Theory of Majorization and its Application*, Academic Press, New York, 1979.
16. H. Minc, Doubly stochastic matrices with minimal permanents, *Pacific J. Math.* **58** (1975), 155–157.
17. H. Minc, *Permanents*, *Encyclopedia of Mathematics and its Applications*, vol. 6, Addison-Wesley, Reading, Mass., 1978.
18. H. Minc, Theory of permanents 1978–1981, *Linear and Multilinear Algebra* **12** (1983), 227–263.
19. H. Minc, The van der Waerden permanent conjecture, *General Inequalities*, vol. 3 (E. F. Beckenbach and W. Walter, eds.), Birkhäuser, Basel, 1983, 23–40.
20. H. Minc, Theory of permanents 1982–1985, *Linear and Multilinear Algebra* **21** (1987), 109–148.
21. L. Mirsky, Results and problems in the theory of doubly-stochastic matrices, *Z. Wahrsch. Verw. Gebiete* **1** (1963), 319–334
22. R. F. Muirhead, Some methods applicable to identities and inequalities of symmetric algebraic functions of n letters, *Proc. Edinburgh Math. Soc.* **21** (1903), 144–157.
23. I. Schur, Über eine Klasse von Mittelbildungen mit Anwendung auf die Determinantentheorie, *Sber. Berliner Math. Ges.* **22** (1923), 9–20.
24. B. L. van der Waerden, Aufgabe 45, *Jber. Deutsch. Math.-Verein.* **35** (1926), 117.

VI

Other Classes of Nonnegative Matrices

6.1. STOCHASTIC MATRICES

A nonnegative square matrix is called *row stochastic*, or simply *stochastic*, if all its row sums are 1. *Column stochastic* matrices are defined similarly.

Stochastic matrices play an important part in the theory of finite homogeneous Markov chains. Let $S_1, S_2, \ldots, S_n$ be n possible states of a certain process or system, and suppose that the probability p_{ij} of the process moving from state i to state j is independent of time for $i, j = 1, 2, \ldots, n$. Then the process is called a *finite homogeneous Markov chain*, and the stochastic $n \times n$ matrix $P = (p_{ij})$ is called the *transition matrix* for the chain. The process is fully determined by its transition matrix, and vice versa.

In this section we study algebraic properties of stochastic matrices. The following properties are immediate consequences of the definition of a stochastic matrix and properties of general nonnegative matrices.

Theorem 1.1. (a) *A nonnegative $n \times n$ matrix A is stochastic if and only if*

$$AJ = J,$$

where J is the $n \times n$ matrix of 1's.

(b) *A square nonnegative matrix A is stochastic if and only if $u = (1, 1, \ldots, 1)$ is an eigenvector corresponding to the maximal eigenvalue* 1 *of A.*

(c) *The moduli of the eigenvalues of a stochastic matrix cannot exceed* 1.

(d) *The product of stochastic matrices is stochastic.*

[For, if A and B are stochastic $n \times n$ matrices, that is, $AJ = BJ = J$, then $(AB)J = A(BJ) = AJ = J$.]

(e) *The set of $n \times n$ stochastic matrices forms a convex polyhedron with the n^n stochastic $n \times n$ $(0,1)$-matrices as its vertices.*

Spectral properties of stochastic matrices do not differ much from those of other nonnegative matrices, particularly irreducible matrices, or any nonnegative matrices with maximal eigenvalue equal to 1 and a corresponding positive eigenvector. In fact, we have the following result.

Theorem 1.2. *If A is a nonnegative matrix with positive maximal eigenvalue r and a positive maximal eigenvector, then $r^{-1}A$ is diagonally similar to a stochastic matrix.*

Proof. Let $x = (x_1, x_2, \ldots, x_n)$ be a positive maximal eigenvector of A. If $D = \operatorname{diag}(x_1, x_2, \ldots, x_n)$, then $x = Du$. Clearly, 1 is the maximal eigenvalue of $D^{-1}(r^{-1}A)D = r^{-1}D^{-1}AD$. Now,

$$\begin{aligned} r^{-1}D^{-1}ADu &= r^{-1}D^{-1}Ax \\ &= r^{-1}D^{-1}rx \\ &= D^{-1}x \\ &= u. \end{aligned}$$

Hence u is a maximal eigenvector of $r^{-1}D^{-1}AD = D^{-1}(r^{-1}A)D$, and it follows from Theorem 1.1(b) that the matrix is stochastic. ∎

The condition in Theorem 1.2 that the matrix have a positive maximal eigenvector is essential. For example, the matrix

$$\begin{bmatrix} 1 & 1 \\ 0 & 1 \end{bmatrix}$$

has maximal eigenvalue 1, but it is not similar to any stochastic matrix. Indeed, a stochastic matrix cannot have an elementary divisor $(\lambda - 1)^i$ with $i > 1$. The analogous result for doubly stochastic matrices is quite obvious. However, it is neither obvious nor easy to prove for stochastic matrices. We require the following auxiliary results. The first of them describes a *normal form* of a reducible matrix.

Lemma 1.1. *A reducible matrix is cogredient to a matrix of the form*

$$M = \begin{bmatrix} A_{11} & 0 & \cdots & 0 & 0 & & \cdots & 0 \\ 0 & A_{22} & \cdots & 0 & 0 & & \cdots & 0 \\ \vdots & & \ddots & & \vdots & & & \vdots \\ 0 & 0 & \cdots & A_{kk} & 0 & & \cdots & 0 \\ A_{k+1,1} & A_{k+1,2} & \cdots & A_{k+1,k} & A_{k+1,k+1} & 0 & \cdots & 0 \\ \vdots & \vdots & & \vdots & & & \ddots & \vdots \\ A_{s1} & A_{s2} & \cdots & A_{sk} & A_{s,k+1} & & \cdots & A_{ss} \end{bmatrix}, \quad (1)$$

where the blocks A_{ij} are $n_i \times n_j$, $i, j = 1, 2, \ldots, s$, the blocks $A_{11}, A_{22}, \ldots, A_{ss}$ are irreducible (possibly 1×1 zero matrices), *and $A_{t1} + A_{t2} + \cdots + A_{t,t-1} \neq 0$, $t = k+1, k+2, \ldots, s$.*

Proof. If A is reducible, then it is cogredient to a matrix of the form

$$\begin{bmatrix} B & 0 \\ C & D \end{bmatrix},$$

where B and D are square submatrices. If both B and D are irreducible, the proof is finished. Otherwise, we can continue the process until we obtain a matrix of the form

$$\begin{bmatrix} A^{11} & 0 & 0 & \cdots & 0 \\ A^{21} & A^{22} & 0 & \cdots & 0 \\ A^{31} & A^{32} & A^{33} & \cdots & 0 \\ \vdots & \vdots & & \ddots & \vdots \\ A^{s1} & A^{s2} & A^{s3} & \cdots & A^{ss} \end{bmatrix}, \tag{2}$$

where the A^{ii} are square irreducible matrices. Suppose that all the off-diagonal blocks in rows $i_1 = 1, i_2, \ldots, i_{k-1}, i_k$ of matrix (2) are 0. By an appropriate permutation of rows and the same permutation of columns, the blocks $A^{i_t i_t}$, $t = 1, 2, \ldots, k$, can be brought to the first k places along the main diagonal. Thus we obtain a matrix of form (1) which is cogredient to the original matrix. ∎

Note that a normal form of a stochastic matrix is stochastic, and so are the blocks $A_{11}, A_{22}, \ldots, A_{kk}$.

The next two lemmas deal with spectral properties of general matrices. We prove them here for completeness.

Lemma 1.2. *Let A be an $n \times n$ matrix, and suppose that*

$$A = \begin{bmatrix} B & 0 \\ C & D \end{bmatrix}, \tag{3}$$

where B is $k \times k$. If B and D have no eigenvalues in common, then A is similar to $B \dotplus D$.

Proof. Let

$$S = \begin{bmatrix} I_k & 0 \\ X & I_{n-k} \end{bmatrix}.$$

We assert that $S^{-1}AS = B \dotplus D$ for an appropriate choice of X. Now,

$$\begin{aligned} S^{-1}AS &= \begin{bmatrix} I_k & 0 \\ -X & I_{n-k} \end{bmatrix} \begin{bmatrix} B & 0 \\ C & D \end{bmatrix} \begin{bmatrix} I_k & 0 \\ X & I_{n-k} \end{bmatrix} \\ &= \begin{bmatrix} B & 0 \\ -XB + C + DX & D \end{bmatrix}, \end{aligned}$$

and the problem is to determine whether there exists an $(n-k)\times k$ matrix X such that $-XB+C+DX=0$, that is,

$$XB - DX = C. \tag{4}$$

We show first that the reduced equation

$$XB - DX = 0 \tag{5}$$

has only the trivial solution $X=0$.

Let H_{m_i} denote the nilpotent $m_i \times m_i$ (0,1)-matrix with 1's in positions $(t, t+1)$, $t = 1,2,\ldots, m_i - 1$, and 0's elsewhere. Let $G = \sum_{i=1}^{p}(\lambda_i I_{m_i} + H_{m_i})$, be the Jordan normal form of B, and $K = \sum_{j=1}^{q}(\mu_j I_{n_j} + H_{n_j})$ be the Jordan normal form of D. Let $G = P^{-1}BP$ and $K = Q^{-1}DQ$. Then equation (5) becomes

$$XPGP^{-1} - QKQ^{-1}X = 0. \tag{6}$$

Now, multiply equation (6) on the left by Q^{-1} and on the right by P. The equation becomes

$$Q^{-1}XPG - KQ^{-1}XP = 0,$$

or

$$YG = KY, \tag{7}$$

where $Y = Q^{-1}XP$. Partition Y into blocks Y_{ij}, so that the block Y_{ij} is $n_i \times m_j$, $i = 1,2,\ldots, q$, $j = 1,2,\ldots, p$. Equating the (i, j) blocks on both sides of (7) we obtain

$$Y_{ij}\left(\lambda_j I_{m_j} + H_{m_j}\right) = \left(\mu_i I_{n_i} + H_{n_i}\right)Y_{ij},$$

that is,

$$\left(\lambda_j - \mu_i\right)Y_{ij} = H_{n_i}Y_{ij} - Y_{ij}H_{m_j}. \tag{8}$$

Now, multiply (8) by $\lambda_j - \mu_i$, and apply equality (8) to the right-hand side of the resulting equation:

$$\begin{aligned}
\left(\lambda_j - \mu_i\right)^2 Y_{ij} &= \left(\lambda_j - \mu_i\right)\left(H_{n_i}Y_{ij} - Y_{ij}H_{m_j}\right) \\
&= H_{n_i}\left(H_{n_i}Y_{ij} - Y_{ij}H_{m_j}\right) - \left(H_{n_i}Y_{ij} - Y_{ij}H_{m_j}\right)H_{m_j} \\
&= H_{n_i}^2 Y_{ij} - 2H_{n_i}Y_{ij}H_{m_j} + Y_{ij}H_{m_j}^2 \\
&= \sum_{\alpha+\beta=2} (-1)^{\beta}\binom{2}{\beta} H_{n_i}^{\alpha} Y_{ij} H_{m_j}^{\beta},
\end{aligned}$$

where α and β are nonnegative integers. Therefore, inductively,

$$(\lambda_j - \mu_i)^\gamma Y_{ij} = (\lambda_j - \mu_i) \sum_{\alpha+\beta=\gamma-1} (-1)^\beta \binom{\gamma-1}{\beta} H_{n_i}^\alpha Y_{ij} H_{m_j}^\beta$$

$$= \sum_{\alpha+\beta=\gamma} (-1)^\beta \binom{\gamma}{\beta} H_{n_i}^\alpha Y_{ij} H_{m_j}^\beta, \tag{9}$$

for any integer $\gamma \geq 2$. If $\gamma = n_i + m_j - 1$, then for any nonnegative integers α and β satisfying $\alpha + \beta = \gamma$, either $\alpha \geq n_i$ or $\beta \geq m_j$, and therefore either $H_{n_i}^\alpha = 0$ or $H_{m_j}^\beta = 0$. It follows that for $\gamma = n_i + m_j - 1$ equation (9) becomes

$$(\lambda_j - \mu_i)^\gamma Y_{ij} = 0.$$

Since $\lambda_j \neq \mu_i$ for all i and j, we must have $Y_{ij} = 0$ for all i and j, and thus $Y = 0$. Hence $X = 0$, and the matrix equation (5) has only the trivial solution $X = 0$. In other words, the corresponding system of $(n-k)k$ linear equations in $(n-k)k$ unknowns x_{ij} has only the trivial solution. We can conclude therefore that the matrix equation (4) has a unique solution for any matrix C. The result now follows. ∎

Lemma 1.3. *Let A be a square matrix, and suppose that*

$$A = \begin{bmatrix} B & 0 \\ C & D \end{bmatrix},$$

where B and D are principal submatrices of A. If λ_t is an eigenvalue of B but is not an eigenvalue of D, then $\lambda I_n - A$ and $\lambda I_h - B$ have exactly the same elementary divisors of the form $(\lambda - \lambda_t)^i$.

Proof. Let A and B be $n \times n$ and $h \times h$, respectively. If B and D have no eigenvalues in common, then, by Lemma 1.2, A is similar to $B \dotplus D$, and the result follows. Otherwise, let $B_1 \dotplus B_2$ be the Jordan normal form of B, where all the main diagonal entries of B_1 are equal to λ_t and none of the main diagonal entries of B_2 is equal to λ_t. Then A is similar to a matrix of the form

$$\begin{bmatrix} B_1 & 0 \\ C' & D' \end{bmatrix},$$

where $D' = B_2 \dotplus D$. Now, B_1 and D' have no eigenvalues in common, and therefore, by Lemma 1.2, A is similar to $B_1 \dotplus D'$. The result follows. ∎

Theorem 1.3. *If A is a stochastic $n \times n$ matrix, then all the elementary divisors of $\lambda I_n - A$ of the form $(\lambda - 1)^i$ are linear.*

Proof. If A is irreducible, then 1 is a simple root of A, and $\lambda - 1$ is the only elementary divisor of $\lambda I_n - A$ involving the eigenvalue 1. If A is reducible, then, by Lemma 1.1, it is cogredient, and therefore similar, to a stochastic matrix M of form (1). Recall that all the main diagonal blocks of M are irreducible. Now, each of the last $n - k$ main diagonal blocks has at least one row sum strictly less than 1, and therefore, by Theorem 1.1, Chapter II, its maximal eigenvalue is strictly less than 1. Let $B = \Sigma_{i=1}^{\cdot k} A_{ii}$ be an $h \times h$ matrix. Then, by Lemma 1.3, $\lambda I_n - A$ and $\lambda I_h - B$ have the same elementary divisors involving the eigenvalue 1. Now, each $I_{n_i} - A_{ii}$, $i = 1, 2, \dots, k$, has exactly one elementary divisor $\lambda - 1$ and no other elementary divisors involving the eigenvalue 1, since $A_{11}, A_{22}, \dots, A_{kk}$ are irreducible and stochastic. It follows that $\lambda I_n - M$, and thus also $\lambda I_n - A$, has k elementary divisors $\lambda - 1$, and no elementary divisors $(\lambda - 1)^i$ with $i > 1$. ■

The final result in this section deals with localization of eigenvalues of a stochastic matrix. It is due to Fréchet [3], but it is really a special case of a theorem of Geršgorin [6] on eigenvalues of general complex matrices.

Theorem 1.4. *If $A = (a_{ij})$ is a row stochastic matrix, and $\omega = \min_i(a_{ii})$, then*

$$|\lambda_t - \omega| \le 1 - \omega,$$

for any eigenvalue λ_t of A.

Proof. Let λ_t be an eigenvalue of an $n \times n$ stochastic matrix A, and let $x = (x_1, x_2, \dots, x_n)$ be a corresponding eigenvector. Let $0 < |x_m| = \max_i(|x_i|)$. Then $\lambda_t x = Ax$, and, in particular,

$$\lambda_t x_m = \sum_{j=1}^{n} a_{mj} x_j,$$

and therefore

$$\lambda_t - a_{mm} = \sum_{j \ne m} a_{mj} (x_j / x_m).$$

Now, by the triangle inequality,

$$\begin{aligned} |\lambda_t - a_{mm}| &\le \sum_{j \ne m} a_{mj} |x_j / x_m| \\ &\le \sum_{j \ne m} a_{mj} \\ &= 1 - a_{mm}, \end{aligned}$$

since A is stochastic. Thus

$$\begin{aligned} |\lambda_t - \omega| &= |\lambda_t - a_{mm} + a_{mm} - \omega| \\ &\le |\lambda_t - a_{mm}| + |a_{mm} - \omega| \\ &\le (1 - a_{mm}) + (a_{mm} - \omega) \\ &= 1 - \omega. \quad \blacksquare \end{aligned}$$

6.2. TOTALLY NONNEGATIVE MATRICES

A particularly interesting class of nonnegative matrices with many applications is the class of totally nonnegative matrices.

Definition 2.1. A real $m \times n$ matrix A is called *totally nonnegative* (*positive*) if all subdeterminants of A, of all orders, are nonnegative (positive).

A symmetric totally nonnegative (positive) matrix is positive semidefinite (definite). Of course, the converse is not true, that is, a nonnegative symmetric positive definite matrix need not be totally nonnegative (see Problem 4). However, as we shall see in our next example, the converse is true if the matrix also happens to be tridiagonal.

Recall that a square matrix $T = (t_{ij})$ is called *tridiagonal* if $t_{ij} = 0$ whenever $|i - j| > 1$. Thus a tridiagonal matrix T has the form

$$T = (t_{ij}) = \begin{bmatrix} b_1 & c_1 & 0 & \cdots & & & 0 \\ a_1 & b_2 & c_2 & 0 & \cdots & & 0 \\ 0 & a_2 & b_3 & c_3 & 0 & \cdots & 0 \\ \vdots & \ddots & \ddots & \ddots & \ddots & & \vdots \\ 0 & \cdots & 0 & a_{n-2} & b_{n-1} & & c_{n-1} \\ 0 & \cdots & & 0 & a_{n-1} & & b_n \end{bmatrix}. \tag{1}$$

We illustrate the definition of a totally nonnegative matrix and some of its consequences by considering first tridiagonal matrices.

Example 2.1. Show that a nonnegative tridiagonal matrix is totally nonnegative if and only if its principal minors are nonnegative.

We first establish a general expression for a subdeterminant of a tridiagonal matrix. We require the following notation. If $\omega = (\omega_1, \omega_2, \ldots, \omega_k)$ is a sequence in $Q_{k,n}$, and g, h are integers, $1 \le g < h \le k$, then $\omega^{(g,h)}$ denotes the subsequence $(\omega_g, \omega_{g+1}, \ldots, \omega_h)$. Now, let $\alpha = (\alpha_1, \alpha_2, \ldots, \alpha_k)$ and $\beta = (\beta_1, \beta_2, \ldots, \beta_k)$ be sequences in $Q_{k,n}$, and let T be the tridiagonal matrix in (1).

We assert that if $\alpha_h \neq \beta_h$ for some h, then

$$\begin{aligned}\det(T[\alpha|\beta]) &= \det\left(T\left[\alpha^{(1,h-1)}|\beta^{(1,h-1)}\right]\right)\det\left(T\left[\alpha^{(h,k)}|\beta^{(h,k)}\right]\right)\\ &= \det\left(T\left[\alpha^{(1,h)}|\beta^{(1,h)}\right]\right)\det\left(T\left[\alpha^{(h+1,k)}|\beta^{(h+1,k)}\right]\right). \quad (2)\end{aligned}$$

If $\alpha_h < \beta_h$, then $\alpha_{h-1} \leq \beta_h - 2$, and therefore $\alpha_i \leq \beta_j - 2$ for $i = 1, 2, \ldots, h-1$ and $j = h, h+1, \ldots, k$. Hence $t_{\alpha_i\beta_j} = 0$ for these values of i and j, and

$$(T[\alpha|\beta])[1, 2, \ldots, h-1|h, h+1, \ldots, k] = 0. \quad (3)$$

On the other hand, if $\alpha_h > \beta_h$, then $\alpha_i \geq \beta_j + 2$ for $i = h, h+1, \ldots, k$ and $j = 1, 2, \ldots, h-1$, and therefore $t_{\alpha_i\beta_j} = 0$ for these values of i and j, and thus in this case

$$(T[\alpha|\beta])[h, h+1, \ldots, k|1, 2, \ldots, h-1] = 0. \quad (4)$$

The first equality in (2) is now implied by either (3) or (4). The second equality in (2) is proved similarly (Problem 5).

If $h = 1$, then the first formula in (2) is to be interpreted as

$$\det(T[\alpha|\beta]) = t_{\alpha_1\beta_1}\det\left(T\left[\alpha^{(2,k)}|\beta^{(2,k)}\right]\right);$$

and if $h = k$, then the second formula in (2) becomes

$$\det(T[\alpha|\beta]) = \det\left(T\left[\alpha^{(1,k-1)}|\beta^{(1,k-1)}\right]\right)t_{\alpha_k\beta_k}.$$

Now, let $\alpha = (\alpha_1, \alpha_2, \ldots, \alpha_k)$ and $\beta = (\beta_1, \beta_2, \ldots, \beta_k)$ be sequences in $Q_{k,n}$, as before, and suppose that $\alpha_i = \beta_i$ for $i = 1, 2, \ldots, g$, and for $i = h+1, h+2, \ldots, k$, where $g < h$, but $\alpha_i \neq \beta_i$ for $i = g+1, g+2, \ldots, h$. Then we can conclude from formulas (2) that

$$\det(T[\alpha|\beta]) = \det\left(T\left[\alpha^{(1,g)}|\beta^{(1,g)}\right]\right)\left(\prod_{i=g+1}^{h} t_{\alpha_i\beta_i}\right)\det\left(T\left[\alpha^{(h+1,k)}|\beta^{(h+1,k)}\right]\right). \quad (5)$$

Note that if $\alpha_i \neq \beta_i$ for $i = 1, 2, \ldots, h$, and $\alpha_i = \beta_i$ for $i = h+1, h+2, \ldots, k$, then formula (5) reads

$$\det(T[\alpha|\beta]) = \left(\prod_{i=1}^{h} t_{\alpha_i\beta_i}\right)\left(\det\left(T\left[\alpha^{(h+1,k)}|\beta^{(h+1,k)}\right]\right)\right);$$

and if $\alpha_i = \beta_i$ for $i = 1, 2, \ldots, g$, and $\alpha_i \neq \beta_i$ for $i = g+1, g+2, \ldots, k$, then formula (5) becomes

$$\det(T[\alpha|\beta]) = \det\left(T\left[\alpha^{(1,g)}|\beta^{(1,g)}\right]\right)\left(\prod_{i=g+1}^{k} t_{\alpha_i\beta_i}\right).$$

It follows from formula (5) that any subdeterminant of a tridiagonal matrix is a product of some of its principal minors and off-diagonal elements. We can conclude therefore that a tridiagonal matrix is totally nonnegative if and only if it is nonnegative and all its principal minors are nonnegative. ■

The necessary and sufficient conditions in Example 2.1 that a nonnegative tridiagonal matrix be totally nonnegative are the same as those that are necessary and sufficient for a hermitian matrix to be positive semidefinite. It is well known that for nonsingular matrices these conditions can be considerably refined. In fact, a hermitian $n \times n$ matrix H is positive definite if merely all its *leading principal minors* are positive [i.e., $\det(H[1,2,\ldots,t|1,2,\ldots,t]) > 0$ for $t = 1,2,\ldots,n$]. It is remarkable that analogous conditions are sufficient for an irreducible tridiagonal matrix to be totally nonnegative.

Example 2.2. Show that an irreducible (nonnegative) tridiagonal matrix is totally nonnegative if and only if all its leading principal minors are positive.

Let $T = (t_{ij})$ be an $n \times n$ irreducible tridiagonal matrix of form (1). If T is totally nonnegative, then, by the result in Example 2.1, all its principal minors (and therefore all its leading principal minors) must be nonnegative. It remains to prove the converse. Since T is irreducible, all the a_i and the c_i must be positive. Note also that if some a_i appears as a factor in a nonzero diagonal product $\prod_j t_{j\sigma(j)}$, then c_i must also be a factor of the same product, and vice versa (see Problem 7). In other words, the numbers a_i and c_i appear in any diagonal product only as products $a_i c_i$. The same is true of any leading principal minor of T. Consider therefore the tridiagonal symmetric matrix

$$T' = \begin{bmatrix} b_1 & d_1 & 0 & \cdots & & & 0 \\ d_1 & b_2 & d_2 & 0 & \cdots & & 0 \\ 0 & d_2 & b_3 & d_3 & 0 & \cdots & 0 \\ \vdots & \ddots & \ddots & \ddots & \ddots & & \vdots \\ 0 & \cdots & 0 & d_{n-2} & b_{n-1} & & d_{n-1} \\ 0 & \cdots & & 0 & d_{n-1} & & b_n \end{bmatrix},$$

where $d_i = \sqrt{a_i c_i}$, $i = 1,2,\ldots,n-1$. By the preceding remarks, every leading principal minor of T is equal to the corresponding minor of T'. Hence all the leading principal minors of T' are positive, and, by properties of symmetric matrices, T' is positive definite. Therefore all principal minors of T' are positive, and it follows that all principal minors of T must be positive, and T is totally nonnegative by the result in Example 2.1. ■

We now return to general totally nonnegative matrices, and start with two of their rather obvious properties.

Theorem 2.1. (a) *The set of $n \times n$ totally nonnegative matrices is closed under multiplication.*

(b) *The product of a totally positive matrix and a nonsingular totally nonnegative matrix is totally positive.*

We leave the proofs of the theorems above as exercises for the reader (Problem 10).

Recall that if $\alpha = (\alpha_1, \alpha_2, \ldots, \alpha_k) \in Q_{k,n}$, then $s(\alpha)$ denotes the sum $\alpha_1 + \alpha_2 + \cdots + \alpha_k$. We require the following preliminary general result relating a minor of the inverse of any matrix to a minor of the matrix.

Lemma 2.1. *If A is a nonsingular $n \times n$ matrix, and $\alpha, \beta \in Q_{r,n}$, then*

$$\det(A^{-1}[\alpha|\beta]) = (-1)^{s(\alpha)+s(\beta)}\det(A(\beta|\alpha))/\det(A).$$

Proof. Let $C_r(M)$ denote the rth compound of M. Then

$$C_r(A)(C_r(A))^{-1} = C_r(A)C_r(A^{-1}) = I,$$

where the identity matrix I is $\binom{n}{r}$-square. In scalar form, the equality reads

$$\sum_{\alpha \in Q_{r,n}} \det(A[\omega|\alpha])\det(A^{-1}[\alpha|\beta]) = \delta_{\omega\beta}, \tag{6}$$

for any ω and β in $Q_{r,n}$. On the other hand, expanding the determinant of A by rows indexed with ω, and using the generalized "Rule of False Cofactors," we obtain

$$\sum_{\alpha \in Q_{r,n}} (-1)^{s(\alpha)+s(\beta)}\det(A[\omega|\alpha])\det(A(\beta|\alpha)) = \delta_{\omega\beta}\det(A). \tag{7}$$

Since the matrix $(C_r(A))^{-1}$ is uniquely determined, we obtain, by comparing (6) and (7),

$$\det(A^{-1}[\alpha|\beta]) = (-1)^{s(\alpha)+s(\beta)}\det(A(\beta|\alpha))/\det(A). \quad ■$$

***Definition* 2.2.** A real matrix A is said to be *sign-regular* if

$$(-1)^{s(\alpha)+s(\beta)}\det(A[\alpha|\beta]) \geq 0, \tag{8}$$

for all α and β in $Q_{k,n}$, $k = 1, 2, \ldots, n$. If all the inequalities (8) are strict, then A is called *strictly sign-regular.*

An immediate consequence of Lemma 2.1 and our definitions is the following theorem.

Theorem 2.2. (a) *A nonsingular matrix A is totally nonnegative if and only if A^{-1} is sign-regular.*

(b) *A nonsingular matrix A is totally positive if and only if A^{-1} is strictly sign-regular.*

This theorem in turn implies the following result.

Theorem 2.3. *Let A be a nonnegative $n \times n$ matrix with a positive determinant. Then A is totally nonnegative if and only if its $(n-1)$th compound is totally nonnegative.*

We omit the proofs of both theorems and relegate them to the problem section (Problems 11 and 12).

Note that Theorem 2.3 is not true for the rth compound of a nonnegative matrix if $2 \le r \le n-2$. For example, it is easy to verify, using the result in Example 2.2, that the tridiagonal matrix

$$T = \begin{bmatrix} 2 & 1 & 0 & 0 \\ 1 & 2 & 1 & 0 \\ 0 & 1 & 2 & 1 \\ 0 & 0 & 1 & 2 \end{bmatrix}$$

is totally nonnegative. However, its second compound

$$C_2(T) = \begin{bmatrix} 3 & 2 & 0 & 1 & 0 & 0 \\ 2 & 4 & 2 & 2 & 1 & 0 \\ 0 & 2 & 4 & 1 & 2 & 0 \\ 1 & 2 & 1 & 3 & 2 & 1 \\ 0 & 1 & 2 & 2 & 4 & 2 \\ 0 & 0 & 0 & 1 & 2 & 3 \end{bmatrix}$$

is not totally nonnegative since it contains several negative minors; e.g., $\det(C_2(T)[1,2|3,4]) = -2$.

A totally nonnegative symmetric matrix is positive semidefinite, and thus it inherits all the properties of positive semidefinite matrices. In particular, the determinant of a positive semidefinite matrix: (a) is zero if the matrix has a zero principal minor; (b) cannot exceed the product of its main diagonal entries (*Hadamard's inequality*); or more generally, cannot exceed the product of one of its principal minors and its complementary minor (*Fisher's in-*

equality). It is remarkable that totally nonnegative matrices have analogous properties even if they do not happen to be symmetric.

We use the following abbreviated notation for submatrices. A principal submatrix $A[\gamma|\gamma]$ is denoted by $A[\gamma]$, and its complementary submatrix $A(\gamma|\gamma)$ by $A(\gamma)$. If $\gamma = (\gamma_1, \gamma_2, \dots, \gamma_s)$ is a sequence in $Q_{s,n}$, and r is an integer, then the sequence $(\gamma_1 + r, \gamma_2 + r, \dots, \gamma_s + r)$ is denoted by $\gamma + r$.

We require the following classical result, known as *Sylvester's identity*.

Lemma 2.2. *Let $A = (a_{ij})$ be any $n \times n$ matrix and let ω denote the sequence $(1, 2, \dots, k)$, where $1 \le k < n$. Define the $(n-k) \times (n-k)$ matrix $B = (b_{ij})$ by*

$$b_{ij} = \det(A[\omega, i+k|\omega, j+k]), \qquad i, j = 1, 2, \dots, n-k.$$

If α and β are sequences in $Q_{t,n-k}$, then

$$\det(B[\alpha|\beta]) = \det(A[\omega])^{t-1}\det(A[\omega, \alpha+k|\omega, \beta+k]). \tag{9}$$

For a proof of this lemma see [4], Chapter II, Section 3.

We also require the following auxiliary result which is of interest by itself.

Lemma 2.3. *If $A = (a_{ij})$ is a totally nonnegative matrix with a zero principal minor, then* $\det(A) = 0$.

Proof. We can assume without loss of generality that the zero principal minor is $\det(A[1, 2, \dots, k])$. If $k = 1$, that is, if $a_{11} = 0$, then $\det(A[1, i|1, j]) = -a_{i1}a_{1j}$, $i, j = 2, 3, \dots, n$. Since the entries and the minors of A are nonnegative, we must have either $a_{i1} = 0$ for $i = 2, 3, \dots, n$, or $a_{1j} = 0$ for $j = 2, 3, \dots, n$. In either case A contains a zero line, and therefore $\det(A) = 0$. Now, assume that $k > 1$ and suppose that $\det(A[1, 2, \dots, t]) > 0$ for $t = 1, 2, \dots, k-1$, and $\det(A[1, 2, \dots, k]) = 0$. Let $B = (b_{ij})$ be the $(n-k+1) \times (n-k+1)$ matrix whose (i, j) entry is

$$b_{ij} = \det(A[1, 2, \dots, k-1, i+k-1|1, 2, \dots, k-1, j+k-1]),$$
$$i, j = 1, 2, \dots, n-k+1.$$

Consider any subdeterminant of B, $\det(B[\alpha|\beta])$, where $\alpha = (\alpha_1, \alpha_2, \dots, \alpha_s)$ and $\beta = (\beta_1, \beta_2, \dots, \beta_s)$ are sequences in $Q_{s,n-k+1}$. By Sylvester's identity,

$$\det(B[\alpha|\beta]) = \det(A[\omega])^{s-1}\det(A[\omega, \alpha+k-1|\omega, \beta+k-1]),$$

where $\omega = (1, 2, \dots, k-1)$. Thus $\det(B[\alpha|\beta])$ is nonnegative for all α and β, and therefore B is totally nonnegative. But $b_{11} = \det(A[1, 2, \dots, k]) = 0$. It follows from the first part of the proof that $\det(B) = 0$. Hence applying

Sylvester's identity again, we have

$$0 = \det(B) = \det(A[1,2,\ldots,k-1])^{n-k}\det(A),$$

which implies that $\det(A) = 0$. ■

Corollary 2.1. *If the determinant of a principal submatrix $A[\omega]$ of a totally nonnegative matrix A is equal to zero, then the determinant of any principal submatrix of A containing $A[\omega]$ as a submatrix also vanishes.*

We next prove the analogue of Fisher's inequality for totally nonnegative matrices.

Theorem 2.4. *If A is a totally nonnegative $n \times n$ matrix and k is an integer, $1 \leq k < n$, then*

$$\det(A) \leq \det(A[1,2,\ldots,k])\det(A(1,2,\ldots,k)).$$

Proof. If $A = (a_{ij})$ has a zero principal minor, then, by Lemma 2.3, $\det(A) = 0$, and the result holds trivially. We can assume therefore that all principal minors of A are positive. We use induction on n. If $n = 2$, then

$$\det(A) = a_{11}a_{22} - a_{12}a_{21} \leq a_{11}a_{22} = \det(A[1|1])\det(A(1|1)).$$

Assume now that $n > 2$, and that the theorem holds for all totally nonnegative $m \times m$ matrices with $m < n$. We can also assume without loss of generality that $k \geq 2$, since the cases $k = 1$ and $k = n - 1$ are analogous. Denote the sequence $(1, 2, \ldots, k-1)$ by ω, and let $B = (b_{ij})$ be again the $(n-k+1) \times (n-k+1)$ matrix whose (i, j) entry is $\det(A[\omega, i+k-1|\omega, j+k-1])$, $i, j = 1, 2, \ldots, n-k+1$. As we saw in the proof of Lemma 2.3, the matrix B is totally nonnegative. Thus

$$\begin{aligned}
\det(A) &= \frac{\det(B)}{\det(A[\omega])^{n-k}} \\
&\leq \frac{b_{11}\det(B(1))}{\det(A[\omega])^{n-k}}, \quad \text{by the induction hypothesis,} \\
&= \frac{b_{11}}{\det(A[\omega])^{n-k}}\det(A[\omega])^{n-k-1}\det(A(k)),
\end{aligned}$$

again by virtue of Sylvester's identity. Now, $b_{11} = \det(A[\omega, k])$, and therefore the above inequality, after some simplification, becomes

$$\det(A) \leq \frac{\det(A[\omega, k])}{\det(A[\omega])}\det(A(k)).$$

Hence applying our induction hypothesis to the determinant of the totally nonnegative $(n-1) \times (n-1)$ matrix $A(k)$, we obtain

$$\det(A) \le \frac{\det(A[\omega, k])}{\det(A[\omega])} \det(A[\omega])\det(A(\omega, k))$$

$$= \det(A[1,2,\ldots,k])\det(A(1,2,\ldots,k)). \quad \blacksquare$$

Theorem 2.4 implies an analogue of Hadamard's inequality for totally nonnegative matrices.

Corollary 2.2. *If $A = (a_{ij})$ is a totally nonnegative $n \times n$ matrix, then*

$$\det(A) \le \prod_{i=1}^{n} a_{ii}.$$

Totally nonnegative matrices have another striking property in common with positive semidefinite symmetric matrices.

Theorem 2.5. *All the eigenvalues of a totally nonnegative matrix are real and nonnegative.*

Proof. Let $\lambda_1, \lambda_2, \ldots, \lambda_n$ be the eigenvalues of a totally nonnegative $n \times n$ matrix A, $\lambda_1 \ge |\lambda_2| \ge |\lambda_3| \ge \cdots \ge |\lambda_n|$. Suppose that $\lambda_i \ne 0$ for $i = 1,2,\ldots,k$, and $\lambda_i = 0$ for $i = k+1, k+2, \ldots, n$. Since A is totally nonnegative, the compound matrix $C_r(A)$ is nonnegative. Hence, by properties of compound matrices ([7], Chapter I, 2.15.12), the maximal eigenvalue of $C_r(A)$ is $\lambda_1\lambda_2 \cdots \lambda_r$ which, by Theorem 4.2, Chapter I, is real and nonnegative. This is true for $r = 1,2,\ldots,k$, and therefore $\lambda_1, \lambda_2, \ldots, \lambda_k$ are positive. $\blacksquare$

6.3. OSCILLATORY MATRICES

The class of totally nonnegative square matrices is perhaps too large to possess noteworthy spectral properties. On the other hand, the class of square totally positive matrices is too restrictive, particularly in applications to the theory of small oscillations. In this section we study the so-called oscillatory matrices which form a subclass of square totally nonnegative matrices that contains totally positive matrices.

Definition 3.1. A totally nonnegative square matrix A is called *oscillatory* if there exists a positive integer m such that A^m is totally positive. The least positive integer k for which A^k is totally positive is called the *exponent* of A.

Some rather obvious properties of oscillatory matrices are listed in the following theorem.

Theorem 3.1. (a) *An oscillatory matrix is nonsingular and primitive.*

(b) *Any positive integer power of an oscillatory matrix is oscillatory.*

(c) *If A is an oscillatory matrix of exponent k, then A^t is totally positive for any integer t, greater than or equal to k.*

(d) *If A is an oscillatory $n \times n$ matrix, then $C_{n-1}(A)$ is also oscillatory* (see Theorem 2.3).

Oscillatory matrices possess truly remarkable spectral properties. We have already shown that all eigenvalues of totally nonnegative matrices are real and nonnegative (Theorem 2.5). For oscillatory matrices we have the following substantially strengthened result.

Theorem 3.2. *The eigenvalues of an oscillatory matrix are all distinct and positive.*

Proof. Let A be an $n \times n$ oscillatory matrix with exponent k, and let $\lambda_1, \lambda_2, \ldots, \lambda_n$, $\lambda_1 > |\lambda_2| \geq \cdots \geq |\lambda_n|$, be the eigenvalues of A. Consider $C_r(A)$, the rth compound of A. Then $C_r(A)$ is nonnegative, and its eigenvalues are

$$\lambda_{\omega_1}\lambda_{\omega_2} \cdots \lambda_{\omega_r}, \qquad (\omega_1, \omega_2, \ldots, \omega_r) \in Q_{r,n}.$$

Since the matrix A^m is totally positive for some positive integer m, the compound matrix $C_r(A^m) = (C_r(A))^m$ is positive, and thus $C_r(A)$ is primitive. This is true for $r = 1, 2, \ldots, n$. It follows that $\lambda_1\lambda_2 \cdots \lambda_r$, the maximal eigenvalue of $C_r(A)$, is positive for $r = 1, 2, \ldots, n$, and therefore $\lambda_1, \lambda_2, \ldots, \lambda_n$ are all real and positive. Hence all the eigenvalues of $C_r(A)$ are real and positive. In particular,

$$\lambda_1\lambda_2 \cdots \lambda_{r-1}\lambda_r > \lambda_1\lambda_2 \cdots \lambda_{r-1}\lambda_{r+1},$$

for $r = 2, 3, \ldots, n-1$, and we can conclude that

$$\lambda_1 > \lambda_2 > \cdots > \lambda_n. \quad \blacksquare$$

The following results concern variations in sign of coordinates in eigenvectors of oscillatory matrices.

Definition 3.2. Let v be a real n-tuple. We count the number of variations in sign in the sequence of the coordinates of v, that is, the number of consecutive coordinates of v with different signs, zero coordinates (if any) being assigned arbitrary signs. The maximum number of sign variations thus obtained is denoted by $S_M(v)$, and their minimum number by $S_m(v)$. If $S_M(v) = S_m(v)$, then v is said to have an *exact* number of sign variations which is denoted by $S(v)$.

Our main theorem is a consequence of the following two lemmas which are of interest in themselves.

Lemma 3.1. *Let A be an oscillatory $n \times n$ matrix with eigenvalues $\lambda_1 > \lambda_2 > \cdots > \lambda_n$, and let $v_1, v_2, \ldots, v_n$ be corresponding eigenvectors, respectively. If w is any nonzero linear combination of vectors $v_1, v_2, \ldots, v_q$, then $S_M(w) \le q - 1$.*

Proof. Since A has distinct eigenvalues, by Theorem 3.2, it is diagonalizable. Let

$$U^{-1}AU = L = \operatorname{diag}(\lambda_1, \lambda_2, \ldots, \lambda_n), \tag{1}$$

where $U^{(j)} = \pm v_j$, $j = 1, 2, \ldots, n$, where the signs are chosen so that $\det(U[1, 2, \ldots, q])$ is positive. Relation (1) implies an analogous relation for compound matrices:

$$C_q(U)^{-1} C_q(A) C_q(U) = C_q(L).$$

Since A is oscillatory, A^m is totally positive for some positive integer m. It follows that $(C_q(A))^m > 0$, and therefore $C_q(A)$ is primitive. Since the first column of $C_q(U)$ is an eigenvector of $C_q(A)$ corresponding to its maximal eigenvalue, the entries in that column must be either all positive or all negative. But these entries are $\det(U[\omega|1, 2, \ldots, q])$, $\omega \in Q_{q,n}$, and since U was constructed so that $\det(U[1, 2, \ldots, q])$ is positive, we have

$$\det(U[\omega|1, 2, \ldots, q]) > 0, \tag{2}$$

for all $\omega \in Q_{q,n}$.

Let

$$w = (w_1, w_2, \ldots, w_n) = \sum_{j=1}^{q} c_j v_j,$$

where the c_j are arbitrary real numbers, not all zero. Suppose that $S_M(w) \ge q$. Then w would have $q + 1$ coordinates, $w_{\alpha_1}, w_{\alpha_2}, \ldots, w_{\alpha_{q+1}}$, where $\alpha = (\alpha_1, \alpha_2, \ldots, \alpha_{q+1}) \in Q_{q+1,n}$, satisfying

$$w_{\alpha_i} w_{\alpha_{i+1}} \le 0, \qquad i = 1, 2, \ldots, q. \tag{3}$$

We note that the w_{α_j} cannot all be zero. For, if they were, then the system

$$\left(U[\alpha_1, \alpha_2, \ldots, \alpha_q|1, 2, \ldots, q]\right)x = 0$$

would have a nontrivial solution $[c_1, c_2, \ldots, c_q]^{\mathrm{T}}$, which would contradict (2).

Let $v_j = (v_{1j}, v_{2j}, \ldots, v_{nj})$, $j = 1, 2, \ldots, n$. Consider the determinant of the $(q+1) \times (q+1)$ matrix

$$\begin{bmatrix} v_{\alpha_1 1} & v_{\alpha_1 2} & \cdots & v_{\alpha_1 q} & w_{\alpha_1} \\ v_{\alpha_2 1} & v_{\alpha_2 2} & \cdots & v_{\alpha_2 q} & w_{\alpha_2} \\ \vdots & \vdots & & \vdots & \vdots \\ v_{\alpha_{q+1} 1} & v_{\alpha_{q+1} 2} & \cdots & v_{\alpha_{q+1} q} & w_{\alpha_{q+1}} \end{bmatrix}.$$

On the one hand, the determinant of the matrix vanishes, since the last column in the matrix is a linear combination of the preceding columns. On the other hand, by expansion on the last column, the determinant of the matrix is equal to

$$\sum_{t=1}^{q+1} (-1)^{t+q+1} w_{\alpha_t} \det\left(U\left[\alpha_1, \ldots, \alpha_{t-1}, \alpha_{t+1}, \ldots, \alpha_{q+1} | 1, 2, \ldots, q\right]\right). \quad (4)$$

But, by (3), the numbers $(-1)^t w_{\alpha_t}$, $t = 1, 2, \ldots, q+1$, have the same sign and are not all zero, whereas, by (2), the minors $\det(U[\alpha_1, \ldots, \alpha_{t-1}, \alpha_{t+1}, \ldots, \alpha_{q+1} | 1, 2, \ldots, q])$ are all positive. Hence the expression (4) cannot vanish. Thus the assumption that $S_M(w) \geq q$ has led to a contradiction, and we must have $S_M(w) \leq q - 1$. ■

Corollary 3.1. *If A is the matrix defined in the statement of Lemma 3.1 and u is a nonzero linear combination of eigenvectors $v_p, v_{p+1}, \ldots, v_q$, then $S_M(u) \leq q - 1$.*

Lemma 3.2. *Let A be an oscillatory $n \times n$ matrix with eigenvalues $\lambda_1 > \lambda_2 > \cdots > \lambda_n$, and let $v_1, v_2, \ldots, v_n$ be eigenvectors corresponding to these eigenvalues, respectively. If w is a nonzero linear combination of vectors $v_p, v_{p+1}, \ldots, v_n$, then*

$$S_m(w) \geq p - 1.$$

Proof. Let $u = \sum_{j=p}^{n} d_j v_j$. We have

$$Av_j = \lambda_j v_j, \qquad j = 1, 2, \ldots, n,$$

and therefore

$$A^{-1} v_j = \lambda_j^{-1} v_j,$$

or

$$(\operatorname{adj} A) v_j = \lambda_j^{-1} v_j \det(A), \qquad j = 1, 2, \ldots, n. \quad (5)$$

Let D be the $n \times n$ generalized permutation matrix with nonzero entries in positions $(i, n-i+1)$, $i = 1, 2, \ldots, n$, the $(i, n-i+1)$ entry being $(-1)^{i+1}$, $i = 1, 2, \ldots, n$. Then $D^{-1} = (-1)^{n+1}D$, and $D(\text{adj } X)D^{-1} = C_{n-1}(X)^{\text{T}}$ for any $n \times n$ matrix X. Now, it follows from (5) that

$$D(\text{adj } A)D^{-1}Dv_j = \lambda_j^{-1}Dv_j\det(A),$$

and therefore

$$C_{n-1}(A)^{\text{T}}(Dv_j) = \lambda_j^{-1}(Dv_j)\det(A), \qquad j = 1, 2, \ldots, n.$$

Thus $Dv_1, Dv_2, \ldots, Dv_n$ are eigenvectors of $C_{n-1}(A)^{\text{T}}$, corresponding to the eigenvalues $\lambda_1^{-1}\det(A), \lambda_2^{-1}\det(A), \ldots, \lambda_n^{-1}\det(A)$, respectively, where $\lambda_n^{-1}\det(A) > \lambda_{n-1}^{-1}\det(A) > \cdots > \lambda_1^{-1}\det(A)$. Now, by Theorem 3.1(d), $C_{n-1}(A)^{\text{T}}$ is oscillatory. Hence Lemma 3.1 applied to this matrix and to the vector $Du = \sum_{j=p}^{n} d_j(Dv_j)$ yields

$$S_M(Du) \leq n - p. \tag{6}$$

Observe that for any real n-tuple v,

$$S_m(v) + S_M(Dv) = n - 1. \tag{7}$$

For, if two adjacent nonzero coordinates of v have the same sign, then the corresponding coordinates of Dv have opposite signs, and if they have different signs, then the corresponding coordinates of Dv have the same sign. If at least one of two adjacent coordinates of v is zero, then this occurrence counts as a change of sign in $S_M(Dv)$ but not in $S_m(v)$. Thus in all cases, every pair of adjacent coordinates gives rise to a change of sign either to be counted in $S_m(v)$ or in $S_M(Dv)$, but not in both. Equality (7) now follows.

From (6) and (7) we have

$$\begin{aligned} S_m(u) &= (n-1) - S_M(Du) \\ &\geq (n-1) - (n-p) \\ &= p - 1. \quad \blacksquare \end{aligned}$$

Corollary 3.2. *If A is an oscillatory $n \times n$ matrix with eigenvalues $\lambda_1 > \lambda_2 > \cdots > \lambda_n$ and corresponding eigenvectors $v_1, v_2, \ldots, v_n$, and if u is a nonzero linear combination of vectors $v_p, v_{p+1}, \ldots, v_q$, then $S_m(u) \geq p - 1$.*

Corollary 3.3. *If A is an oscillatory $n \times n$ matrix with eigenvalues $\lambda_1 > \lambda_2 > \cdots > \lambda_n$ and corresponding eigenvectors $v_1, v_2, \ldots, v_n$, and if u is a nonzero linear combination of vectors $v_p, v_{p+1}, \ldots, v_q$, then*

$$p - 1 \leq S_m(u) \leq S_M(u) \leq q - 1.$$

The following theorem is the main result in this sequence of lemmas and corollaries. It is obtained from Corollary 3.3 by setting $p = q$.

Theorem 3.3. *If A is an oscillatory $n \times n$ matrix with eigenvalues $\lambda_1 > \lambda_2 > \cdots > \lambda_n$ and corresponding eigenvectors $v_1, v_2, \ldots, v_n$, then*

$$S(v_j) = j - 1, \qquad j = 1, 2, \ldots, n.$$

We conclude the section with a result giving simple necessary and sufficient conditions for a totally nonnegative matrix to be oscillatory. We quote the result without proof. For a proof see [5], Chapter II, Theorem 10.

Theorem 3.4. *A totally nonnegative $n \times n$ matrix $A = (a_{ij})$ is oscillatory if and only if it is nonsingular, and*

$$a_{i,i+1} > 0, \qquad a_{i+1,i} > 0,$$

for $i = 1, 2, \ldots, n - 1$.

6.4. *M*-MATRICES

In this section we introduce the so-called M-matrices. Although these matrices are not nonnegative, they are closely related to nonnegative matrices. For example, as we shall see, the inverse of a nonsingular M-matrix is always nonnegative. Their main raison d'être, however, are their applications in mathematics and economics. Here we study some of their algebraic properties.

Of many possible definitions of M-matrices the one given below stresses an important connection between this class of matrices and nonnegative matrices.

Definition 4.1. A real $n \times n$ matrix A is an *M-matrix* if there exists a nonnegative matrix B with maximal eigenvalue r such that

$$A = cI_n - B, \tag{1}$$

where $c \geq r$. Note that the main diagonal entries of an M-matrix are nonnegative, and all its other entries are nonpositive. The set of $n \times n$ real matrices whose off-diagonal entries are nonpositive is denoted by F_n.

There are many other definitions of M-matrices. In fact, Berman and Plemmons [1] list 50 equivalent definitions for nonsingular M-matrices! We state and prove a few of the alternative necessary and sufficient conditions for a matrix in F_n to be an M-matrix, that are relevant to the main theme of this book.

Theorem 4.1. *A matrix in F_n is an M-matrix if and only if all its eigenvalues have a nonnegative real part.*

Proof. Let $A = (a_{ij}) \in F_n$, and suppose that every eigenvalue of A has a nonnegative real part. Let $a_{mm} = \max_i(a_{ii})$. Then $B = a_{mm}I_n - A$ is nonnegative. Let r be the maximal eigenvalue of B. Now $A = a_{mm}I_n - B$, and therefore $a_{mm} - r$ is a real eigenvalue of A. It follows that $a_{mm} - r$ is nonnegative, that is, $a_{mm} \geq r$. Thus $A = a_{mm}I_n - B$ is an M-matrix.

Now, let $A = cI_n - B$ be an M-matrix, and let r be the maximal eigenvalue of B, $c \geq r$. Let λ_t be an eigenvalue of A. Suppose that $\mathrm{Re}(\lambda_t)$, the real part of λ_t, is negative. Then

$$\begin{aligned} 0 &= \det(\lambda_t I_n - A) \\ &= \det(\lambda_t I_n - cI_n + B) \\ &= \det((c - \lambda_t)I_n - B), \end{aligned}$$

and thus $c - \lambda_t$ is an eigenvalue of B. But c is nonnegative and $-\mathrm{Re}(\lambda_t)$ is assumed to be positive, and therefore

$$\begin{aligned} |c - \lambda_t| &\geq c - \mathrm{Re}(\lambda_t) \\ &> c \\ &\geq r, \end{aligned}$$

which contradicts the maximality of r. ∎

In the following theorem the sufficiency of the condition in Theorem 4.1 is substantially weakened by restricting it to the real eigenvalues of the matrix.

Theorem 4.2. *A matrix A in F_n is an M-matrix if and only if every real eigenvalue of A is nonnegative.*

Proof. The necessity of the condition follows directly from Theorem 4.1. Suppose now that all real eigenvalues of $A \in F_n$ are nonnegative. Let $B = a_{mm}I_n - A$, where $a_{mm} = \max_i a_{ii}$. Then B is a nonnegative matrix. Let r be its maximal eigenvalue. Since $a_{mm} - r$ is a real eigenvalue of $A = a_{mm}I_n - B$, it must be nonnegative. In other words, $a_{mm} \geq r$, and therefore A is an M-matrix. ∎

The next characterization of M-matrices involves their principal minors.

Theorem 4.3. *A principal submatrix of an M-matrix is an M-matrix.*

Proof. Let A be an M-matrix, and let $A = cI_n - B$, where B is a nonnegative matrix with maximal eigenvalue r, and $c \geq r$. Let $\omega \in Q_{k,m}$, and let μ be the maximal eigenvalue of the principal submatrix $B[\omega]$. Then, by Theorem 5.2, Chapter I, $r \geq \mu$. Thus $c \geq \mu$, and therefore

$$A[\omega] = cI_k - B[\omega]$$

is an M-matrix. ■

From Theorem 4.3 we can deduce the following result.

Theorem 4.4. *A matrix in F_n is an M-matrix if and only if all its principal minors are nonnegative.*

Proof. In order to prove the necessity of the condition it suffices, by virtue of Theorem 4.3, to show that the determinant of an M-matrix is nonnegative. By Theorem 4.2, the real roots of an M-matrix are nonnegative. Moreover, since the matrix is real, all its nonreal eigenvalues occur in conjugate pairs. It follows that the determinant of an M-matrix, which is the product of its eigenvalues, is nonnegative.

Now suppose that $A \in F_n$, and all its principal minors are nonnegative. Let $a_{mm} = \max_i a_{ii}$, and let $B = a_{mm}I_n - A$. Then $A = a_{mm}I_n - B$, where B is a nonnegative matrix with maximal eigenvalue r, say. If all the principal minors of A are zero, then all its eigenvalues are zero, and $A = -B$ since $a_{mm} = 0$. In this case $r = 0$, and therefore $a_{mm} \geq r = 0$, and it follows that A is an M-matrix. We assume now that all principal minors of A are nonnegative but not all are zero. Let $E_i(A)$ denote the ith elementary symmetric function of the eigenvalues of A, that is, the sum of all its principal minors of order i. Then $E_i(A) \geq 0$ for all i, and the inequality is strict for at least one i. If p is a positive number, then

$$\begin{aligned}\det((p + a_{mm})I_n - B) &= \det(pI_n + (a_{mm}I_n - B)) \\ &= \det(pI_n + A) \\ &= \sum_{k=0}^{n} p^{n-k}E_{n-k}(A),\end{aligned}$$

which is positive. Hence $p + a_{mm}$ cannot be an eigenvalue of B for any positive number p. Thus $a_{mm} \geq r$, and therefore $A = a_{mm}I_n - B$ is an M-matrix. ■

The next result is one of the most important and relevant to our study of nonnegative matrices.

Theorem 4.5. *A nonsingular matrix $A \in F_n$ is an M-matrix if and only if A^{-1} is nonnegative.*

Proof. Let $A = cI_n - B$ be a nonsingular M-matrix, where B is a nonnegative matrix with maximal eigenvalue $r < c$. Note that r cannot be equal to c since A is nonsingular. Thus the moduli of the eigenvalues of B/c are all less than 1. It follows that $\lim_{t\to\infty}(B/c)^t = 0$ (see Problem 1, Chapter III). Now,

$$(I_n - B/c)\big(I_n + B/c + (B/c)^2 + \cdots (B/c)^p\big) = I_n - (B/c)^{p+1}.$$

Hence

$$(I_n - B/c)\sum_{t=0}^{\infty}(B/c)^t = I_n,$$

and therefore

$$(I_n - B/c)^{-1} = \sum_{t=0}^{\infty}(B/c)^t,$$

which is nonnegative since $\sum_{t=0}^{p}(B/c)^t$ is nonnegative for every positive integer p. But

$$A^{-1} = c^{-1}(I_n - B/c)^{-1}$$

which, as we have shown above, is nonnegative.

Suppose now that A is a matrix in F_n and $A^{-1} \geq 0$. Let $a_{mm} = \max_i a_{ii}$, and $B = a_{mm}I_n - A$. Then B is a nonnegative matrix. Let r be its maximal root, and let x be a nonnegative eigenvector, so that $Bx = rx$. We have

$$\begin{aligned} Ax &= (a_{mm}I_n - B)x \\ &= (a_{mm} - r)x, \end{aligned}$$

and therefore

$$x = (a_{mm} - r)A^{-1}x.$$

But x and $A^{-1}x$ are nonnegative, and $a_{mm} - r \neq 0$, since A is nonsingular. Hence $a_{mm} - r > 0$, and $A = a_{mm}I_n - B$ is an M-matrix. ■

One of the attractive features of the theory of M-matrices is the abundance of equivalent definitions of M-matrices involving concepts from many areas of linear algebra, and the relative ease with which most of them can be proved. Several of them, connected with nonnegative matrices, are listed in the problem section below. We conclude the section with the following example.

Example 4.1. Let A be a singular M-matrix, $A = cI_n - B$, where B is an irreducible matrix with maximal eigenvalue r, $c \geq r$. Show that A has rank $n - 1$.

Since A is singular, c is an eigenvalue of $cI_n - A = B$. Hence $c \leq r$, by maximality of r. Thus we have $c = r$. But r is a simple eigenvalue of B, and therefore the eigenvalue 0 of $A = rI_n - B$ must be simple as well. It follows that the rank of A is $n - 1$. ■

PROBLEMS

1 Let

$$A = \begin{bmatrix} 3 & 2 & 0 \\ 1 & 1 & 1 \\ 0 & 1 & 3 \end{bmatrix},$$

and let r be the maximal eigenvalue of A. Find a stochastic matrix similar to $r^{-1}A$.

2 Let S be the stochastic matrix constructed in Problem 1. Find the eigenvalues of S, and use them to verify Theorem 1.4.

3 Prove that a totally nonnegative symmetric matrix is positive semidefinite.

4 Find a positive definite symmetric matrix with real positive entries that is not totally nonnegative.

5 Prove the second equality in (2), Section 6.2.

6 Let $\alpha = (1,2,4,5,7,8,9)$ and $\beta = (1,3,4,6,7,8,9)$, and let $T = (t_{ij})$ be an $n \times n$ tridiagonal matrix, $n \geq 9$. Express the determinant of $T[\alpha|\beta]$ as a product of principal minors of T and some of its off-diagonal entries.

7 Let $T = (t_{ij})$ be an irreducible tridiagonal matrix of form (1), Section 6.2. Show that if some a_i is a factor in a nonzero diagonal product $\prod_j t_{j\sigma(j)}$, then c_i must be also a factor in the product.

8 Let T be a positive semidefinite symmetric tridiagonal matrix. Show that T is totally nonnegative if and only if all its entries are nonnegative.

9 Show that a generalized Vandermonde matrix $V = (r_i^{\alpha_j})$, $0 < r_1 < r_2 < \cdots < r_n$ and $\alpha_1 < \alpha_2 < \cdots < \alpha_n$, is totally positive. [*Hint*: First prove (or assume) that the determinant of a generalized Vandermonde matrix is always positive.]

10 Prove both parts of Theorem 2.1.

11 Prove in detail both parts of Theorem 2.2.

12 Prove Theorem 2.3.

13 Show that the second compound matrix of a totally positive 4×4 matrix need not be totally nonnegative.

14 Give a counterexample to each of the following false assertions.

(a) A primitive totally nonnegative matrix is oscillatory.

(b) If A is a nonnegative matrix and A^k is totally positive, then A is oscillatory.

(c) A totally nonnegative matrix with distinct positive eigenvalues is oscillatory.

15 Let A be the matrix in Problem 1. Is A oscillatory?

16 Let $B = A + I_3$, where A is the matrix in Problem 1. Show by direct computation, that B is oscillatory.

17 Determine necessary and sufficient conditions for a real n-tuple to have an exact number of sign variations (see Definition 3.1).

18 Let B be the matrix in Problem 16. Find eigenvectors v_1, v_2, v_3 of B corresponding to its eigenvalues $\lambda_1, \lambda_2, \lambda_3$, where $\lambda_1 > \lambda_2 > \lambda_3$, and verify that $S(v_1) = 0$, $S(v_2) = 1$, and $S(v_3) = 2$.

19 For what values of t is the matrix

$$A(t) = \begin{bmatrix} 3 & -2 & 1 \\ -1 & 2 & -2 \\ -2 & 0 & t \end{bmatrix}$$

an M-matrix?

20 If $A(t)$ is the matrix defined in the preceding problem, find a lower bound for the moduli of the eigenvalues of $A(5)$ and of $A(6)$. (*Hint*: Use Theorem 1.1, Chapter II.)

21 Show that all triangular matrices in F_n with nonnegative main diagonal entries are M-matrices.

22 Show that the result in Example 4.1 is not valid if the matrix B is reducible.

23 Let A be an $n \times n$ M-matrix. Show that $A + D$ is an M-matrix for any nonnegative diagonal matrix D.

24 Let $A \in F_n$. Show that A is an M-matrix if and only if $A + D$ is nonsingular for every positive diagonal matrix D.

25 Let A be an M-matrix. Show that there exists a nonnegative n-tuple x such that Ax is nonnegative.

26 Let A be an M-matrix, $A = cI_n - B$, where B is an irreducible nonnegative matrix with maximal eigenvalue r, $c \geq r$. Show that there exists a positive n-tuple y such that $Ay \geq 0$. Show also that if B is reducible, then there may not exist a positive n-tuple y for which $Ay \geq 0$.

27 Show that a matrix $A \in F_n$ is an M-matrix if and only if every real eigenvalue of each principal submatrix of A is nonnegative.

28 A matrix in F_n is said to be a *Stieltjes matrix* if it is positive definite symmetric. Show that a Stieltjes matrix is an M-matrix.

REFERENCES

1. A. Berman and R. J. Plemmons, *Nonnegative Matrices in the Mathematical Sciences*, Academic Press, New York, 1979.
2. M. Fiedler and V. Ptak, On matrices with non-positive off-diagonal elements and positive principal minors, *Czechoslovak Math. J.* **12** (1962), 382–400.
3. M. Fréchet, Comportement asymptotique des solutions d'un système d'équations linéaires et homogènes aux différences finies du premier ordre à coefficients constants, *Publ. Fac. Sci. Univ. Masaryk*, No. 178 (1933), 1–24.
4. F. R. Gantmacher, *The Theory of Matrices*, Chelsea, New York, 1959.
5. F. R. Gantmakher and M. G. Krein, *Oszillationsmatrizen, Oszillationskerne, und kleine Schwingungen mechanischer Systeme*, Akademic-Verlag, Berlin, 1960.
6. S. A. Geršgorin, Über die Abgrenzung der Eigenwerte einer Matrix, *Izv. Akad. Nauk, SSSR, Ser. Fiz-Mat.* **6** (1931), 749–754.
7. M. Marcus and H. Minc, *A Survey of Matrix Theory and Matrix Inequalities*, Allyn and Bacon, Boston, 1964.
8. A. Ostrowski, Über die Determinanten mit überwiegender Hauptdiagonale, *Comment. Math. Helv.* **10** (1937), 69–96.
9. G. Poole and T. Boullion, A survey on M-matrices, *SIAM Rev.* **16** (1974), 419–427.

VII

Inverse Eigenvalue Problems

7.1. INVERSE EIGENVALUE PROBLEMS FOR NONNEGATIVE MATRICES AND STOCHASTIC MATRICES

In this chapter we consider the following inverse problems involving nonnegative matrices.

1. *The inverse eigenvalue problem.* Determine necessary and sufficient conditions for a complex number to be an eigenvalue of a nonnegative matrix (or a stochastic or doubly stochastic matrix).
2. *The inverse spectrum problem.* Determine necessary and sufficient conditions for a complex n-tuple σ to be the spectrum of a nonnegative $n \times n$ matrix.
3. *The inverse elementary divisor problem.* Determine necessary and sufficient conditions for a complex matrix to be similar to a nonnegative (or a doubly stochastic) matrix.

Obviously, the first of these problems, which is the topic of this section, is the easiest of the three. However, the general problem as set above, without any restrictions on the matrix, is not of much interest. In fact, *any* complex number is an eigenvalue of some nonnegative matrices.

Example 1.1. Find a nonnegative matrix one of whose eigenvalues is $\alpha + \beta i$, where α and β are any real numbers.

Let

$$A(\alpha, \beta) = \begin{bmatrix} |\alpha| & |\beta| & 0 & 0 \\ 0 & |\alpha| & |\beta| & 0 \\ 0 & 0 & |\alpha| & |\beta| \\ |\beta| & 0 & 0 & |\alpha| \end{bmatrix}.$$

It is easily seen that the eigenvalues of $A(\alpha, \beta)$ are

$$|\alpha| \pm |\beta|, \qquad |\alpha| \pm |\beta| i.$$

Now let $B(\alpha, \beta)$ be the 8×8 nonnegative matrix

$$\begin{bmatrix} 0 & A(\alpha, \beta) \\ A(\alpha, \beta) & 0 \end{bmatrix}.$$

Then it follows from Theorem 4.8, Chapter III (or otherwise), that the eigenvalues of $B(\alpha, \beta)$ are

$$\pm|\alpha| \pm |\beta| \quad \text{and} \quad \pm|\alpha| \pm |\beta| i.$$

Clearly, one of the last four eigenvalues is equal to $\alpha + \beta i$ whatever the signs of α and β happen to be. ■

The nonnegative 8×8 matrix constructed in Example 1.1 is by no means the smallest matrix with $\alpha + \beta i$ as one of its eigenvalues. Indeed, it can be shown that any complex number is an eigenvalue of a positive 3×3 circulant (Problem 14).

In order that the inverse eigenvalue problem be meaningful, it is necessary to restrict it to certain classes of nonnegative matrices: e.g., nonnegative $n \times n$ matrices with a prescribed maximal eigenvalue, or $n \times n$ stochastic matrices, or $n \times n$ doubly stochastic matrices, or $n \times n$ nonnegative circulants. It may appear that the inverse eigenvalue problem for nonnegative $n \times n$ matrices is more general than that for $n \times n$ stochastic matrices, but, in fact, the two problems are virtually equivalent (see Theorem 1.2 below). We begin with some simple aspects of the inverse eigenvalue problem for $n \times n$ stochastic matrices that involve convex sums.

Recall that if $z_1, z_2, \ldots, z_k$ are complex numbers, that is, points in the complex plane, and $c_1, c_2, \ldots, c_k$ are nonnegative numbers adding up to 1, then the point (complex number) $\sum_{t=1}^{k} c_t z_t$ is called a *convex sum* of $z_1, z_2, \ldots, z_k$. The set of all convex sums of finite subsets of a set X is called the *convex hull* of X, and is denoted by $H(X)$. The convex hull of a finite set $\{z_1, z_2, \ldots, z_m\}$ is called the *convex polygon spanned by* $z_1, z_2, \ldots, z_m$. In every convex polygon there are unique points, called *vertices* of the polygon, which span the polygon, and have the property that none of them is a convex sum of the others ([9], Chapter II, 1.5.1).

Theorem 1.1. *If α is an eigenvalue of a stochastic $n \times n$ matrix, and ζ is a convex sum of* $1, \alpha, \alpha^2, \ldots, \alpha^k$, *then ζ is an eigenvalue of an $n \times n$ stochastic matrix.*

Proof. Let α be an eigenvalue of a stochastic $n \times n$ matrix A, and let $\zeta = \sum_{t=0}^{k} c_t \alpha^t$. Then ζ is an eigenvalue of the matrix $B = c_0 I_n + c_1 A + c_2 A^2 + \cdots + c_k A^k$. Now, it is easily seen that any nonnegative power of a stochastic matrix is stochastic, and that any convex sum of stochastic matrices is stochastic. The first assertion follows from Theorem 1.1(d), Chapter VI, and the second can be shown to be true by part (a) of the same theorem:

$$\left(\sum_{t=0}^{k} c_t A^t\right) J = \sum_{t=0}^{k} c_t A^t J$$
$$= \sum_{t=0}^{k} c_t J$$
$$= J.$$

Thus B is a stochastic matrix with eigenvalue ζ. ■

Corollary 1.1. *If α is an eigenvalue of a stochastic $n \times n$ matrix, then every point in $H(1, \alpha, \alpha^2, \ldots, \alpha^k, \ldots)$, the convex hull of $1, \alpha, \alpha^2, \ldots, \alpha^k, \ldots,$ is an eigenvalue of an $n \times n$ stochastic matrix.*

Corollary 1.2. *If α is an eigenvalue of a stochastic $n \times n$ matrix, then $\theta\alpha$ is also an eigenvalue of a stochastic $n \times n$ matrix, for any nonnegative number θ not exceeding 1.*

We are now ready to show that the inverse eigenvalue problems for $n \times n$ nonnegative matrices and for $n \times n$ stochastic matrices are equivalent. If the number in question is zero, then it is an eigenvalue of any singular nonnegative matrix and of any singular stochastic matrix. Otherwise, we have the following result.

Theorem 1.2. *A complex nonzero number α is an eigenvalue of a nonnegative $n \times n$ matrix with a positive maximal eigenvalue r, if and only if α/r is an eigenvalue of a stochastic $n \times n$ matrix.*

Proof. The sufficiency of the condition is obvious. Suppose that α is an eigenvalue of a nonnegative $n \times n$ matrix with maximal eigenvalue $r > 0$. If A is irreducible, then, by virtue of Theorem 1.2, Chapter VI, $r^{-1}A$ is similar to a stochastic matrix which clearly has α/r as one of its eigenvalues. If A is reducible, then, by Lemma 1.1, Chapter VI, it is cogredient to a matrix in canonical form [see (1), Section 6.1]. In this case, α is an eigenvalue of one of the main diagonal blocks. Let it be A_{tt}, an $m \times m$ irreducible submatrix with maximal eigenvalue ρ which cannot exceed the maximal eigenvalue r of the matrix. Then A_{tt} is similar to a stochastic $m \times m$ matrix B with eigenvalue α/ρ. Thus α/ρ is an eigenvalue of the stochastic $n \times n$ matrix $B \dotplus I_{n-m}$. But $\rho \leq r$, and therefore, by Corollary 1.2, α/r is an eigenvalue of some stochastic $n \times n$ matrix. ■

In view of the preceding theorem we can confine our discussion to the inverse eigenvalue problem for stochastic matrices. The following result due to Kingman [12] deals with eigenvalues of doubly stochastic matrices.

Theorem 1.3. *Let Π_k denote the set of points in the complex plane bounded by the regular k-sided polygon, $k \geq 2$, inscribed in the unit circle and with one of its vertices at $(0,1)$. Then each of the points in*

$$\Pi^n = \Pi_2 \cup \Pi_3 \cup \cdots \cup \Pi_n$$

is an eigenvalue of a doubly stochastic $n \times n$ matrix.

Proof. Let z be a point in Π_k. Then z is a convex sum of the vertices of Π_k:

$$z = \sum_{t=0}^{k-1} c_t e^{2\pi i t/k},$$

where $c_t \geq 0$ and $\sum_{t=0}^{k-1} c_t = 1$. But $e^{2\pi i/k}$, is an eigenvalue of P_k, the $k \times k$ permutation matrix with 1's in positions $(t, t+1)$, $t = 1, 2, \ldots, k-1$, and $(k, 1)$. Let Q be the $n \times n$ permutation matrix $P_k \dotplus I_{n-k}$. Then z is an eigenvalue of the doubly stochastic $n \times n$ matrix

$$\sum_{t=0}^{k-1} c_t Q^t.$$

Hence every point in Π_k is an eigenvalue of a doubly stochastic $n \times n$ matrix for $k = 2, 3, \ldots, n$. ■

Corollary 1.3. *A complex number is an eigenvalue of a doubly stochastic $n \times n$ circulant if and only if it belongs to Π_n.*

Proof. The eigenvalues of an $n \times n$ circulant C with first row

$$[c_0 \quad c_1 \quad \cdots \quad c_{n-1}]$$

are

$$\sum_{t=0}^{n-1} c_t e^{2\pi i t k/n}, \qquad k = 1, 2, \ldots, n$$

([9], Chapter I, 4.9). If C happens to be doubly stochastic, then $c_0, c_1, \ldots, c_{n-1}$ are nonnegative and $\sum_{t=0}^{n-1} c_t = 1$. Hence every eigenvalue of a doubly stochastic $n \times n$ circulant is a convex sum of vertices of Π_n, and therefore belongs to Π_n. Conversely, if α belongs to Π_n, then

$$\alpha = \sum_{t=0}^{n-1} d_t e^{2\pi i t/n},$$

where the d_t are nonnegative numbers adding up to 1, and therefore α is an eigenvalue of the doubly stochastic circulant whose first line is

$$[d_0 \quad d_1 \quad \cdots \quad d_{n-1}]. \quad \blacksquare$$

It is not known whether any eigenvalues of doubly stochastic $n \times n$ matrices lie outside Π^n (and, of course, inside the unit circle). On the other hand, it is known that eigenvalues of some stochastic matrices do not belong to Π^n [3, 7].

Example 1.2. Find a doubly stochastic 3×3 matrix with eigenvalue $\alpha = (-3 + \sqrt{3}\,i)/12$.

Since α does not lie in Π_2 we try to express it as a convex sum of the vertices of Π_3:

$$(-3 + \sqrt{3}\,i)/12 = c_1 + c_2(-1 + \sqrt{3}\,i)/2 + c_3(-1 - \sqrt{3}\,i)/2, \qquad (1)$$

where c_1, c_2, and c_3 are nonnegative, and

$$c_1 + c_2 + c_3 = 1. \qquad (2)$$

Equality (1) yields

$$4c_1 - 2c_2 - 2c_3 = -1,$$

and

$$6c_2 - 6c_3 = 1,$$

which together with (2) give

$$c_1 = \tfrac{1}{6}, \qquad c_2 = \tfrac{1}{2}, \quad \text{and} \quad c_3 = \tfrac{1}{3}.$$

Thus α lies in Π_3, and it is an eigenvalue of the doubly stochastic matrix

$$\frac{1}{6}I_3 + \frac{1}{2}P + \frac{1}{3}P^2 = \frac{1}{6}\begin{bmatrix} 1 & 3 & 2 \\ 2 & 1 & 3 \\ 3 & 2 & 1 \end{bmatrix}. \quad \blacksquare$$

Example 1.3. Let $\beta = (1 + 3i)/12$. (a) Find a doubly stochastic 4×4 matrix, by expressing β as a convex sum of the vertices of Π_3. (b) Find doubly stochastic 4×4 circulants with eigenvalue β.

(a) Let

$$c_1 + c_2\left(-\frac{1}{2} + \frac{\sqrt{3}}{2}i\right) + c_3\left(-\frac{1}{2} - \frac{\sqrt{3}}{2}i\right) = \frac{1}{12} + \frac{3}{12}i, \qquad (3)$$

where c_1, c_2, and c_3 are nonnegative numbers satisfying

$$c_1 + c_2 + c_3 = 1. \tag{4}$$

Solving equations (3) and (4) for c_1, c_2, and c_3, we obtain

$$c_1 = \frac{7}{18}, \qquad c_2 = \frac{11 + 3\sqrt{3}}{36}, \quad \text{and} \quad c_3 = \frac{11 - 3\sqrt{3}}{36}.$$

Thus the doubly stochastic matrix

$$\frac{1}{36}\begin{bmatrix} 14 & 11 + 3\sqrt{3} & 11 - 3\sqrt{3} & 0 \\ 11 - 3\sqrt{3} & 14 & 11 + 3\sqrt{3} & 0 \\ 11 + 3\sqrt{3} & 11 - 3\sqrt{3} & 14 & 0 \\ 0 & 0 & 0 & 36 \end{bmatrix}$$

has eigenvalue β.

(b) The vertices of Π_4 are $1, i, -1, -i$. Let d_1, d_2, d_3, and d_4 be nonnegative numbers such that

$$d_1 + d_2 i - d_3 - d_4 i = (1 + 3i)/12, \tag{5}$$

and

$$d_1 + d_2 + d_3 + d_4 = 1. \tag{6}$$

Solving equations (5) and (6) we obtain

$$d_1 = \frac{5 - s}{12}, \qquad d_2 = \frac{3 + s}{12}, \qquad d_3 = \frac{4 - s}{12}, \qquad d_4 = \frac{s}{12},$$

where s is a real number satisfying

$$0 \leq s \leq 4, \tag{7}$$

so that the numbers d_1, d_2, d_3, d_4 are nonnegative. Hence β is an eigenvalue of every doubly stochastic circulant

$$\frac{1}{12}\begin{bmatrix} 5 - s & 3 + s & 4 - s & s \\ s & 5 - s & 3 + s & 4 - s \\ 4 - s & s & 5 - s & 3 + s \\ 3 + s & 4 - s & s & 5 - s \end{bmatrix},$$

where s is an arbitrary real number satisfying (7). ■

The reader should note that the method used in the proof of Theorem 1.3 and in the above examples cannot be expected, in general, to produce all $n \times n$ doubly stochastic matrices with a given eigenvalue in Π_k.

Let Θ_n denote the set of points in the complex plane that are eigenvalues of stochastic $n \times n$ matrices. It is easy to see that Θ_n is a compact set. Hence, by Corollary 1.2, in order to determine Θ_n it suffices to find its boundary points, that is, points $\alpha \in \Theta_n$ for which $\rho\alpha \notin \Theta_n$ for any $\rho > 1$. We first determine the points of Θ_n on the unit circle.

Theorem 1.4. *A complex number z with modulus* 1 *is an eigenvalue of an $n \times n$ stochastic matrix if and only if*

$$z = e^{2\pi ik/h}, \tag{8}$$

where h and k are integers satisfying $0 \leq k < h \leq n$.

Proof. Suppose that z is an eigenvalue of a stochastic matrix A, and $|z| = 1$. If A is irreducible, then, by Theorem 1.1, Chapter III, z is an hth root of unity, where $h \leq n$ is the index of imprimitivity of A. If A is reducible, then z is an eigenvalue of B, an irreducible principal $t \times t$ submatrix of A. Thus the maximal eigenvalue of B must be 1, and, by Theorem 1.1, Chapter II, B must be stochastic. It follows that z is an sth root of unity, where s is the index of imprimitivity of B, $s \leq t \leq n$.

Now, suppose that z is a number in (8). If $h = 1$, then $z = 1$, which indeed is an eigenvalue of every stochastic matrix. If $h > 1$, then z is an eigenvalue of the $n \times n$ permutation matrix $P_h \dotplus I_{n-h}$, where P_h is the $h \times h$ permutation matrix with 1's in the superdiagonal and in the $(h, 1)$ position. ∎

Corollary 1.4. *A complex number z of modulus* 1 *is an eigenvalue of an $n \times n$ doubly stochastic matrix if and only if z is one of the numbers in* (8).

Note that $\bigcup_{n=2}^{\infty}\Pi^n$ contains all interior points in the unit circle. Hence Theorem 1.3 and Corollary 1.4 imply the following result.

Theorem 1.5. *A complex number z is an eigenvalue of a doubly stochastic matrix* (*or a stochastic matrix*) *if and only if either it is an interior point of the unit circle, or*

$$z = e^{2\pi ip/q}, \tag{9}$$

where p and q are any integers, $q \neq 0$.

The determination of Θ_n for a given n is more difficult. This problem was partly solved by Dmitriev and Dynkin [2, 3] who used rather complicated geometric methods. We give here two of their results.

If Z is a convex polygon with vertices $z_1, z_2, \ldots, z_k$, and α is a complex number, then αZ denotes the convex polygon spanned by $\alpha z_1, \alpha z_2, \ldots, \alpha z_k$. In what follows we assume that the convex polygon Z is nontrivial, that is, $Z \neq \{0\}$.

Theorem 1.6. *A complex number α is an eigenvalue of an $n \times n$ stochastic matrix if and only if there exists a convex polygon Z with k vertices, $k \leq n$, such that $\alpha Z \subset Z$.*

Proof. Let $A = (a_{ij})$ be an $n \times n$ stochastic matrix, and suppose that $Az = \alpha z$, where $z = (z_1, z_2, \ldots, z_n) \neq 0$. Then

$$\alpha z_i = \sum_{j=1}^{n} a_{ij} z_j, \qquad i = 1, 2, \ldots, n.$$

Thus αz_i belongs to $Z = H(z_1, z_2, \ldots, z_n)$ for $i = 1, 2, \ldots, n$. Hence

$$\alpha Z \subset Z.$$

Conversely, if $Z = H(z_1, z_2, \ldots, z_n)$, and α is a complex number such that αZ is a subpolygon of Z, then $\alpha z_i \in Z$ for $i = 1, 2, \ldots, n$. Therefore each αz_i is a convex sum of the z_i:

$$\alpha z_i = \sum_{j=1}^{n} b_{ij} z_j, \qquad i = 1, 2, \ldots, n, \tag{10}$$

where the b_{ij} are nonnegative numbers satisfying $\sum_{j=1}^{n} b_{ij} = 1$, $i = 1, 2, \ldots, n$. Equalities (10) state that α is an eigenvalue of the $n \times n$ stochastic matrix $B = (b_{ij})$. ∎

Theorem 1.6 can be used to prove Corollary 1.2 and Theorem 1.4 without the use of the results obtained in the first three chapters. It can also be used to establish directly the fact that the modulus of an eigenvalue of a stochastic matrix cannot exceed 1. We leave these proofs as an exercise for the reader (Problems 5, 9, and 10).

We need the following geometric lemma.

Lemma 1.1. *Let Z be a convex k-sided polygon with vertices $z_1, z_2, \ldots, z_k$. Let 0 be an interior point of Z, and let φ_j denote the angle $\angle(z_{j+1}, z_j, 0)$, $j = 1, 2, \ldots, k$ (z_{k+1} being z_1). Then*

$$\min(\varphi_1, \varphi_2, \ldots, \varphi_k) \leq \mu = \frac{\pi}{2} - \frac{\pi}{k}.$$

Note that if Z is a regular k-sided polygon and 0 is its center, then $\varphi_j = \mu$, $j = 1, 2, \ldots, k$.

Proof. Suppose to the contrary that $\varphi_j > \mu$, $j = 1, 2, \ldots, k$. Consider the triangle $\Delta(z_j, z_{j+1}, 0)$.

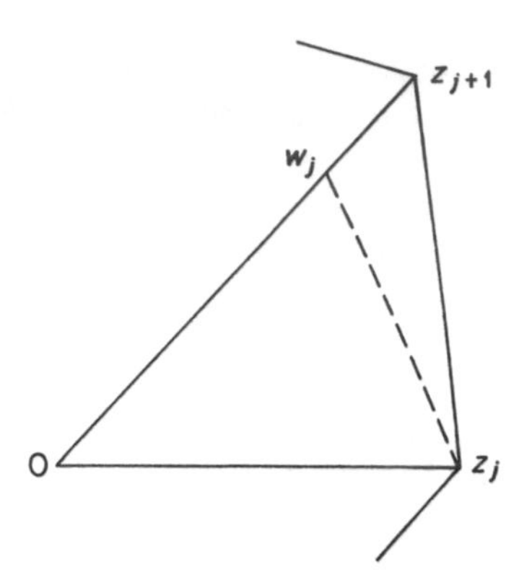

Since $\varphi_j > \mu$, there is a point w_j on the side $0z_{j+1}$ such that the angle $\angle(w_j, z_j, 0)$ is equal to μ. Denote the angle $\angle(0, w_j, z_j)$ by ψ_j. Then in triangle $\Delta(0, z_j, w_j)$ we have

$$\frac{\sin \psi_j}{\sin \mu} = \frac{0z_j}{0w_j} > \frac{0z_j}{0z_{j+1}}, \qquad j = 1, 2, \ldots, k. \tag{11}$$

Multiply inequalities (11) for $j = 1, 2, \ldots, k$,

$$\begin{aligned} \frac{\sin \psi_1}{\sin \mu} \frac{\sin \psi_2}{\sin \mu} \cdots \frac{\sin \psi_k}{\sin \mu} &> \frac{0z_1}{0z_2} \frac{0z_2}{0z_3} \cdots \frac{0z_k}{0z_1} \\ &= 1. \end{aligned} \tag{12}$$

Now, $\psi_j = \pi - \mu - \angle(z_{j+1}, 0, z_j)$. Therefore

$$\begin{aligned} \sum_{j=1}^{k} \psi_j &= k(\pi - \mu) - \sum_{j=1}^{k} \angle(z_{j+1}, 0, z_j) \\ &= k\left(\pi - \frac{\pi}{2} + \frac{\pi}{k}\right) - 2\pi \\ &= \frac{k\pi}{2} - \pi \\ &= k\mu. \end{aligned}$$

But all the ψ_j's are in the interval $(0, \pi)$, and therefore the product $\prod_{j=1}^{k} \sin \psi_j$ attains its maximum when the ψ_j's are equal. Hence $\prod_{j=1}^{k} \sin \psi_j \leq \sin^k \mu$, contradicting (12). ∎

Dmitriev and Dynkin [2] made use of Lemma 1.1 to obtain the following interesting result.

Theorem 1.7. *If the amplitude of a point α in the complex plane does not exceed $2\pi/n$ in absolute value, $n \geq 3$, then α is an eigenvalue of an $n \times n$ stochastic matrix if and only if α lies either in the triangle $\Delta(0, 1, e^{2\pi i/n})$ or in $\Delta(0, 1, e^{-2\pi i/n})$.*

Theorem 1.7 asserts that an $n \times n$ stochastic matrix cannot have eigenvalues inside either of the two segments of the unit circle joining the point 1 with the points $e^{2\pi i/n}$ and $e^{-2\pi i/n}$, respectively.

Proof. If α lies in either of the two triangles, then, by Theorem 1.3, it is an eigenvalue of an $n \times n$ stochastic matrix. Conversely, let $\alpha \in \Theta_n$, and $|\text{amp}(\alpha)| < 2\pi/n$. We show that α lies in one of the two triangles.

Suppose that $0 \leq \text{amp}(\alpha) < 2\pi/n$. By Theorem 1.6, there exists a k-sided convex polygon Z, $k \leq n$, with vertices $z_1, z_2, \ldots, z_k$, such that $\alpha Z \subset Z$. Denote the angle $\angle(z_{j+1}, z_j, 0)$ by φ_j, $j = 1, 2, \ldots, k$. By Lemma 1.1, at least one of these angles, φ_t say, does not exceed $\pi/2 - \pi/k \leq \pi/2 - \pi/n$. The point $\beta = \alpha z_t$ lies inside the polygon Z or on its boundary, and therefore

$$\begin{aligned}\angle(\beta, z_t, 0) &\leq \angle(z_{t+1}, z_t, 0)\\ &\leq \frac{\pi}{2} - \frac{\pi}{n}.\end{aligned}$$

Clearly, the angles $\angle(\beta, z_t, 0)$ and $\angle(\alpha, 1, 0)$ are equal, and therefore $\angle(\alpha, 1, 0) \leq \pi/2 - \pi/n$, and α lies within the triangle $\Delta(0, 1, e^{2\pi i/n})$.

The case $-2\pi/n < \text{amp}(\alpha) \leq 0$ is proved similarly. ■

Example 1.4. Show that the set Θ_3 of eigenvalues of 3×3 stochastic matrices consists of the points in the interior and on the boundary of the triangle with vertices 1, $e^{2\pi i/3}$, and $e^{-2\pi i/3}$, and on the segment $[1, -1]$.

Theorem 1.3 implies that the region described in the statement of the corollary is contained in Θ_3. By Theorem 1.4, no other points on the unit circle are in Θ_3. It remains to show that no nonreal points inside the unit circle but outside the triangle $\Delta(1, e^{2\pi i/3}, e^{-2\pi i/3})$ can be eigenvalues of a 3×3 stochastic matrix. By Theorem 1.7, no points inside the two segments of the unit circle joining the point 1 with the points $e^{2\pi i/3}$ and $e^{-2\pi i/3}$, respectively, are in Θ_3. Now, consider any nonreal point $x + yi$ inside the segment of the unit circle joining the points $e^{2\pi i/3}$ and $e^{-2\pi i/3}$: $-1 < x < -\frac{1}{2}$, $x^2 + y^2 < 1$, $y \neq 0$. If $x + yi$ were an eigenvalue of a 3×3 stochastic matrix, then the other two eigenvalues would have to be 1 and $x - yi$. But then the trace of the matrix would be $1 + 2x < 0$, which is clearly impossible. ■

Dmitriev and Dynkin [3] call a convex polygon Q in the complex plane *cyclic* if it is a convex hull of points $1, \alpha, \alpha^2, \ldots, \alpha^k, \ldots,$ for some number α. Now, the product of two points in Q is clearly in Q, and therefore if $\mu \in Q$, then $\mu Q \subset Q$. It follows from Theorem 1.6 that Θ_n, the set of eigenvalues of

$n \times n$ stochastic matrices, contains all k-sided cyclic polygons, $k \leq n$. Dmitriev and Dynkin conjectured that Θ_n is just the union of all these cyclic polygons, and they proved the conjecture for $n \leq 5$.

The inverse eigenvalue problem for $n \times n$ stochastic matrices was completely solved by Karpelevič [7] who used a different, less restrictive, definition of cyclic polygons:

An n-sided convex polygon Q is *cyclic* if there exists a complex number λ and an integer $p \leq n$ such that Q coincides with the convex hull of points $\lambda^m e^{2\pi i q/p}$, $m = 0, 1, 2, \ldots$; $q = 0, 1, \ldots, p - 1$.

Karpelevič showed in [7] that Θ_n is the union of all thus defined k-sided cyclic polygons, $k \leq n$. He also showed that Θ_n is a curvilinear polygon with vertices as determined in Theorem 1.4.

Theorem 1.8. *The region Θ_n is symmetric relative to the real axis. It is contained within the circle $|z| \leq 1$, and intersects its boundary, $|z| = 1$, at points $e^{2\pi i a/b}$, where a and b run over all integers satisfying $0 \leq a < b \leq n$. The boundary of Θ_n consists of these points and of curvilinear arcs connecting them in circular order. For $n > 3$ each of these arcs is given by one of the following parametric equations*:

$$\lambda^q(\lambda^p - t)^r = (1 - t)^r, \tag{13}$$

$$(\lambda^b - t)^d = (1 - t)^d \lambda^q, \tag{14}$$

where the real parameter t runs over the interval $0 \leq t \leq 1$, and b, d, p, q, r are natural numbers defined as follows.

Consider an arc connecting consecutive points of Θ_n on the unit circle. Let its endpoints be $e^{2\pi i a'/b'}$ and $e^{2\pi i a''/b''}$, in counterclockwise order. Then either

$$b''\left[\frac{n}{b''}\right] \geq b'\left[\frac{n}{b'}\right], \tag{15}$$

or

$$b''\left[\frac{n}{b''}\right] \leq b'\left[\frac{n}{b'}\right]. \tag{16}$$

If (15) *holds for some arc, then* (16) *must hold for its complex conjugate arc. Thus, due to the symmetry of Θ_n, it will suffice to describe arcs satisfying* (15).

Let $r_1 = b''$ and $r_2 = a''$, and let $r_3, r_4, \ldots, r_m$ be the sequence of positive remainders obtained by the repeated use of the Euclidean algorithm: $r_t = r_{t+1}q_t + r_{t+2}$, $0 < r_{t+2} < r_{t+1}$, $t = 1, 2, \ldots, m - 2$, $r_{m-1} = r_m q_{m-1}$. *If* $[n/b''] = 1$ *and $r_{2s} = 1$, for some integer s, then the arc connecting the points $e^{2\pi i a'/b'}$ and $e^{2\pi i a''/b''}$ is given by equation* (13), *where $r = r_{2s-1}$, and p and q are defined by the relations*:

$$a''p \equiv 1 \quad \text{mod } b'', \qquad 0 < p < b'', \tag{17}$$

$$a''q \equiv -r \quad \text{mod } b'', \qquad 0 \leq q < b''. \tag{18}$$

Otherwise, the arc connecting points $e^{2\pi ia'/b'}$ *and* $e^{2\pi ia''/b''}$ *is given by equation* (14), *where* $d = [n/b'']$, $b = b''$, *and* q *is defined by the relation*

$$a''q \equiv -1 \mod b'', \qquad 0 < q < b''. \tag{19}$$

The proof of Theorem 1.8 is too long and too involved to be included here. We refer the reader to Karpelevič's paper [7]. The statement of the theorem, however, although rather lengthy, is not nearly as complicated as some writers claim. In fact, the theorem gives quite a straightforward prescription for constructing Θ_n. We illustrate the method in the following examples.

Example 1.5. Determine the boundary of Θ_4.

The only points of unit modulus in Θ_4 are 1, $e^{\pi i/2}$, $e^{2\pi i/3}$, $e^{\pi i}$, $e^{4\pi i/3}$, and $e^{3\pi i/2}$. By Theorem 1.7, the parts of the boundary of Θ_4 connecting the point 1 with $e^{\pi i/2}$ and $e^{3\pi i/2}$, respectively, are straight lines.

Consider the arc connecting $e^{2\pi i/4}$ with $e^{2\pi i/3}$. Here $a' = 1$, $b' = 4$, $a'' = 1$, and $b'' = 3$, and therefore

$$b''\left[\frac{n}{b''}\right] = 3 < b'\left[\frac{n}{b'}\right] = 4.$$

In order to follow the method described in the statement of Theorem 1.8, we shift our attention to the convex conjugate arc of the one considered above, that is, the arc connecting $e^{4\pi i/3}$ to $e^{3\pi i/2}$. We have now $a' = 2$, $b' = 3$, $a'' = 3$, and $b'' = 4$, and therefore

$$b''\left[\frac{n}{b''}\right] = 4 > b'\left[\frac{n}{b'}\right] = 3,$$

as expected. Since $[n/b''] = 1$, we proceed to compute the sequence of remainders: $r_1 = 4$, $r_2 = 3$, $r_3 = r_1 - r_2q_1 = 1$, and thus $m = 3$. But $r_{2s} = r_2 = 3 \neq 1$, and therefore the applicable parametric equation is (14). The constants are: $d = [n/b''] = 1$, $b = b'' = 4$, and q which is defined by $3q \equiv -1 \mod 4$, $0 < q < 4$, is equal to 1. Hence the parametric equation is

$$(\lambda^4 - t) - (1 - t)\lambda = 0. \tag{20}$$

Note that $\lambda - 1$ is always a factor of (13) and (14). Thus after dividing (20) by $\lambda - 1$ and simplifying the resulting equation we obtain our parametric equation,

$$\lambda^3 + \lambda^2 + \lambda + t = 0. \tag{21}$$

Thus the points of the required arc and its complex conjugate are the nonreal roots of equations obtained from (21) by letting t vary from 0 to 1.

Lastly, we determine the arc connecting $e^{2\pi i/3}$ and $e^{\pi i}$. Here $a' = 1$, $b' = 3$, $a'' = 1$, and $b'' = 2$. This time

$$b''\left[\frac{n}{b''}\right] = 4 > b'\left[\frac{n}{b'}\right] = 3.$$

Since $[n/b''] = 2 \neq 1$, the required parametric equation is derived once more from (14) setting $d = 2$, $b = 2$, and $q \equiv -1 \bmod 2$, that is, $q = 1$. We have

$$(\lambda^2 - t)^2 - (1 - t)^2\lambda = 0,$$

which after division by $\lambda - 1$ and simplification becomes

$$\lambda^3 + \lambda^2 - (2t - 1)\lambda - t^2 = 0. \tag{22}$$

Parametric equation (22) determines the points of the required arc and its complex conjugate. The region Θ_4 is shown in the following figure [17].

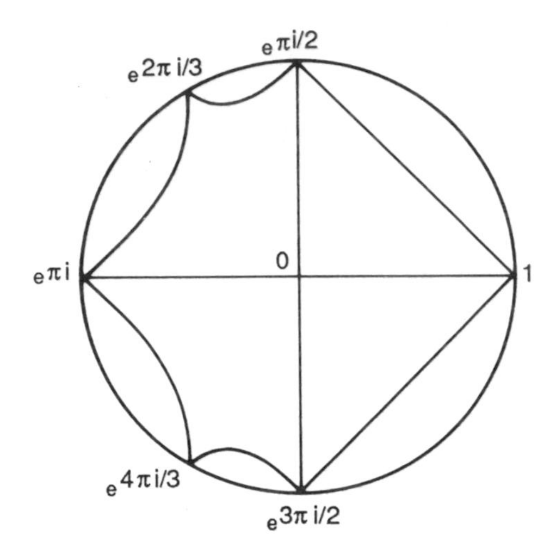

■

Example 1.6. Find the parametric equation of the arc in the boundary of Θ_7 between the points $e^{\pi i}$ and $e^{8\pi i/7}$.

Here our constants are $a' = 1$, $b' = 2$, $a'' = 4$, and $b'' = 7$. Therefore

$$b''\left[\frac{n}{b''}\right] = 7 > b'\left[\frac{n}{b'}\right] = 6.$$

Since $[n/b''] = 1$, we compute the sequence of remainders: $r_1 = 7$, $r_2 = 4$, $r_3 = 7 - 4q_1 = 3$, $r_4 = 4 - 3q_2 = 1$. Now, $r_{2s} = 1$ for $s = 2$, and therefore the required arc is given by parametric equation (13), where $r = r_3 = 3$, and

p,q are given by the relations

$$4p \equiv 1 \quad \mod 7, \qquad 0 < p < 7,$$
$$4q \equiv -3 \quad \mod 7, \qquad 0 \le q < 7,$$

which yield $p = 2$ and $q = 1$. Thus the parametric equation of the arc of the boundary of Θ_7 connecting the points $e^{\pi i}$ and $e^{8\pi i/7}$ is

$$\lambda(\lambda^2 - t)^3 - (1 - t)^3 = 0. \quad \blacksquare$$

7.2. INVERSE SPECTRUM PROBLEMS FOR NONNEGATIVE MATRICES

The problem considered in this section is the determination of necessary and sufficient conditions for an n-tuple of complex numbers $\sigma = (\lambda_1, \lambda_2, \dots, \lambda_n)$ to be the spectrum of a nonnegative matrix. There are two obvious necessary conditions: One is a well-known property of eigenvalues of real matrices, and the other condition is implied by Theorem 4.2, Chapter I,

$$\bar{\sigma} = (\bar{\lambda}_1, \bar{\lambda}_2, \dots, \bar{\lambda}_n) = \sigma, \tag{1}$$

$$\max_i (|\lambda_i|) \in \sigma. \tag{2}$$

Another necessary condition expresses the obvious requisite that the trace of any positive integer power of a nonnegative matrix must be nonnegative:

$$s_m(\sigma) = \sum_{i=1}^{n} \lambda_i^m \ge 0, \qquad m = 1, 2, \dots . \tag{3}$$

Conditions (1) and (2) alone are sufficient in case $n = 2$ (see Problem 17). For $n \ge 3$, however, these two conditions, even with condition (3), are not sufficient. For example, the 5-tuple $\sigma = (1, 1, -\frac{2}{3}, -\frac{2}{3}, -\frac{2}{3})$ satisfies all three conditions, but it cannot be the spectrum of any 5×5 nonnegative matrix, since the matrix would have to be reducible, and thus it would have to contain two complementary principal submatrices, one with eigenvalues $1, -\frac{2}{3}$, and the other with eigenvalues $1, -\frac{2}{3}, -\frac{2}{3}$, which is clearly impossible.

Loewy and London [8] established the following additional necessary condition for a complex n-tuple to be the spectrum of a nonnegative matrix.

Theorem 2.1. *Let $\sigma = (\lambda_1, \lambda_2, \dots, \lambda_n)$ be the spectrum of a nonnegative $n \times n$ matrix. Then*

$$(s_k(\sigma))^m \le n^{m-1} s_{km}(\sigma), \tag{4}$$

for any positive integers k and m.

Proof. Let $A = (a_{ij})$ be the nonnegative $n \times n$ matrix with spectrum σ. Let $B = (b_{ij}) = A^k$, and let $D = \operatorname{diag}(b_{11}, b_{22}, \ldots, b_{nn})$. Then clearly $B^m - D^m$ is nonnegative (Problem 16), and therefore

$$\begin{aligned}\operatorname{tr}(B^m) &\geq \operatorname{tr}(D^m)\\ &= \sum_{i=1}^{n} b_{ii}^m\\ &\geq \frac{1}{n^{m-1}}\left(\sum_{i=1}^{n} b_{ii}\right)^m,\end{aligned}$$

by Hölder's inequality (see [9], Chapter II, 3.3). But

$$\operatorname{tr}(B^m) = \operatorname{tr}(A^{km}) = s_{km}(\sigma),$$

and

$$\sum_{i=1}^{n} b_{ii} = \operatorname{tr}(B) = \operatorname{tr}(A^k) = s_k(\sigma).$$

Hence

$$n^{m-1}s_{km}(\sigma) \geq (s_k(\sigma))^m. \quad \blacksquare$$

Note that for $k = 1$, condition (4) becomes

$$s_m(\sigma) \geq (s_1(\sigma))^m/n^{m-1}, \qquad m = 1, 2, \ldots .$$

Thus (4), together with the condition $s_1(\sigma) \geq 0$, imply condition (3). However, the following example shows that condition (4) is independent of conditions (1), (2), and (3).

Example 2.1 (Loewy and London [8]). Show that condition (4) is independent of conditions (1), (2), and (3).

Let $\sigma = (\sqrt{2}, i, -i)$. Then conditions (1) and (2) are clearly satisfied. Also, $s_1(\sigma) = \sqrt{2}$, $s_2(\sigma) = 0$, and $s_k(\sigma) > 0$ for $k \geq 3$. Thus condition (3) is satisfied. However, if we take $k = 1$ and $m = 2$, then

$$(s_k(\sigma))^m = 2 > n^{m-1}s_{km}(\sigma) = 0,$$

and condition (4) does not hold. $\blacksquare$

Example 2.2 (Loewy and London [8]). Show that conditions (1), (2), (3), and (4) are not sufficient for an n-tuple to be the spectrum of a nonnegative matrix if $n \geq 4$.

Let $\sigma = (\sqrt{2}, \sqrt{2}, i, -i, 0, 0, \ldots, 0)$. Then conditions (1), (2), and (3) are clearly satisfied. The proof that (4) is also satisfied is straightforward.

If $m = 1$, then (4) is an equality. Assume then that $m \geq 2$. If k is an odd integer, then

$$\begin{aligned}(s_k(\sigma))^m &= (2^{1+k/2})^m \\ &\leq 2^{2m-2+km/2} \\ &\leq 4^{m-1}(2^{1+km/2} - 2) \\ &\leq n^{m-1}s_{km}(\sigma),\end{aligned}$$

since $n \geq 4$.

If $k \equiv 2 \bmod 4$, then

$$(s_k(\sigma))^m = (2^{1+k/2} - 2)^m,$$

and (4) holds a fortiori.

Lastly, if $k \equiv 0 \bmod 4$, then

$$\begin{aligned}(s_k(\sigma))^m &= (2^{1+k/2} + 2)^m \\ &\leq (2^{(k+3)/2})^m, \quad \text{since } k \geq 4, \\ &\leq 2^{m(k+4)/2-1}, \quad \text{since } m \geq 2, \\ &= 4^{m-1}(2^{1+km/2}) \\ &\leq n^{m-1}s_{km}(\sigma),\end{aligned}$$

since $km \equiv 0 \bmod 4$. Thus σ satisfies condition (4) as well. However, σ cannot be the spectrum of any nonnegative matrix. For, by virtue of Theorem 4.3, Chapter I, the matrix would have to be reducible, and the nonzero eigenvalues of one of its principal submatrices would have to be $\sqrt{2}$, i, and $-i$. But as we have seen in Example 2.1, these numbers do not satisfy condition (4), and therefore no nonnegative matrix can have these eigenvalues. ■

We now show that for 3×3 matrices even weaker conditions than (1), (2), (3), and (4) are sufficient.

Theorem 2.2 (Loewy and London [8]). *If* $\sigma = (\lambda_1, \lambda_2, \lambda_3)$ *and*

$$\begin{gathered}\bar{\sigma} = \sigma, \\ \lambda_1 \geq |\lambda_j|, \qquad j = 2, 3, \\ \lambda_1 + \lambda_2 + \lambda_3 \geq 0, \\ (\lambda_1 + \lambda_2 + \lambda_3)^2 \leq 3(\lambda_1^2 + \lambda_2^2 + \lambda_3^2),\end{gathered}$$

then there exists a nonnegative matrix with spectrum σ.

Proof. We first show that if σ is a real triple satisfying just the conditions $\lambda_1 \geq |\lambda_j|$, $j = 2, 3$, and $\lambda_1 + \lambda_2 + \lambda_3 \geq 0$, then σ is the spectrum of a nonnegative matrix. Indeed, if either $\lambda_2 \geq 0$ or $\lambda_3 \geq 0$, say $\lambda_2 \geq 0$, then the nonnegative matrix $\lambda_2 \dotplus ((\lambda_1 - \lambda_3)J_2 + \lambda_3 I_2)$ has the required spectrum. If λ_2 and λ_3 are both negative,

$$\frac{1}{2}\begin{bmatrix} \lambda_1 + \lambda_2 + \lambda_3 & \lambda_1 - \lambda_2 + \lambda_3 & 1 \\ \lambda_1 - \lambda_2 + \lambda_3 & \lambda_1 + \lambda_2 + \lambda_3 & 1 \\ -2\lambda_1\lambda_3 & -2\lambda_1\lambda_3 & 0 \end{bmatrix} \tag{5}$$

is nonnegative and has spectrum σ. (See also the companion matrix constructed in the proof of Theorem 2.3 below.)

Now, suppose that σ contains nonreal numbers, and that all the conditions in the statement of the theorem are satisfied. We construct a 3×3 matrix with eigenvalues $\lambda_1, \lambda_2, \lambda_3$. Let $\sigma' = (\rho, e^{i\theta}, e^{-i\theta})$, where $\rho = \lambda_1 / |\lambda_2|$, $\lambda_2 = \overline{\lambda}_3 = |\lambda_2| e^{i\theta}$, $0 < \theta < \pi$. Then the matrix

$$A' = \begin{bmatrix} \rho & 0 & 0 \\ 0 & \cos\theta & \sin\theta \\ 0 & -\sin\theta & \cos\theta \end{bmatrix}$$

has spectrum σ'. We find a nonnegative matrix similar to A'. Let J be the 3×3 matrix of 1's, and let

$$Q = \begin{bmatrix} \frac{1}{\sqrt{3}} & \frac{1}{\sqrt{2}} & \frac{1}{\sqrt{6}} \\ \frac{1}{\sqrt{3}} & -\frac{1}{\sqrt{2}} & \frac{1}{\sqrt{6}} \\ \frac{1}{\sqrt{3}} & 0 & -\frac{2}{\sqrt{6}} \end{bmatrix}.$$

Then $Q^{\mathrm{T}} J Q = \operatorname{diag}(3, 0, 0)$, and therefore

$$\begin{aligned} A'' &= QA'Q^{\mathrm{T}} \\ &= Q(\operatorname{diag}(\rho, 0, 0))Q^{\mathrm{T}} + Q\left(0 \dotplus \begin{bmatrix} \cos\theta & \sin\theta \\ -\sin\theta & \cos\theta \end{bmatrix}\right)Q^{\mathrm{T}} \\ &= \frac{1}{3}\begin{bmatrix} \rho + 2\cos\theta & \rho - 2\cos(\frac{1}{3}\pi + \theta) & \rho - 2\cos(\frac{1}{3}\pi - \theta) \\ \rho - 2\cos(\frac{1}{3}\pi - \theta) & \rho + 2\cos\theta & \rho - 2\cos(\frac{1}{3}\pi + \theta) \\ \rho - 2\cos(\frac{1}{3}\pi + \theta) & \rho - 2\cos(\frac{1}{3}\pi - \theta) & \rho + 2\cos\theta \end{bmatrix}. \end{aligned}$$

Now, the conditions in the statement of the theorem imply that

$$\rho + 2\cos\theta \geq 0, \tag{6}$$

and

$$(\rho + 2\cos\theta)^2 \leq 3(\rho^2 + 2\cos 2\theta),$$

which yield

$$(\rho - 2\cos(\tfrac{1}{3}\pi + \theta))(\rho - 2\cos(\tfrac{1}{3}\pi - \theta)) \geq 0.$$

Thus

$$\rho - 2\cos(\tfrac{1}{3}\pi - \theta) \geq 0, \tag{7}$$

since

$$\rho - 2\cos(\tfrac{1}{3}\pi + \theta) > 0, \tag{8}$$

for $0 < \theta < \pi$ and $\rho \geq 1$. Now, inequalities (6), (7), and (8) imply that the matrix A'' is nonnegative. Also, A'' is similar to A', and therefore σ' is also the spectrum of A''. It follows that σ is the spectrum of the nonnegative matrix $A = |\lambda_2|A''$. ∎

The general inverse spectrum problem for nonnegative matrices is unsolved. The only direction in which some progress has been made is the problem of the existence of a nonnegative matrix with a prescribed real spectrum. We conclude the section with the first, and perhaps the most important, result in this area due to Suleimanova [16] followed by some improvements due to Perfect [14] who also devised the simple proof given here.

Theorem 2.3 (Suleimanova [16]). *Let $\sigma = (\lambda_1, \lambda_2, \ldots, \lambda_n)$ be a real n-tuple satisfying*

$$\lambda_1 + \lambda_2 + \cdots + \lambda_n \geq 0,$$
$$\lambda_j < 0, \qquad j = 2, 3, \ldots, n.$$

Then there exists a nonnegative $n \times n$ matrix with spectrum σ.

Proof. Let

$$f(\lambda) = \prod_{j=1}^{n} (\lambda - \lambda_j)$$
$$= \lambda^n - \sum_{t=1}^{n} c_t \lambda^{n-t}.$$

We assert that all $c_2, c_3, \ldots, c_n$ must be positive. For, the polynomial

$$f(-\lambda) = (-1)^n \left(\lambda^n + \sum_{t=1}^{n} (-1)^{t-1} c_t \lambda^{n-t} \right)$$

has exactly $n-1$ positive roots. Therefore, by Descartes' rule of signs, the number of variations of sign in the sequence

$$1, c_1, -c_2, c_3, -c_4, \ldots, (-1)^{n-1}c_n$$

is exactly $n-1$. However, $c_1 = \lambda_1 + \lambda_2 + \cdots + \lambda_n$ is nonnegative, and it follows that all the other c_t must be positive. Hence the companion matrix of $f(\lambda)$,

$$\begin{bmatrix} 0 & 1 & 0 & \cdots & 0 & 0 \\ 0 & 0 & 1 & \cdots & 0 & 0 \\ \vdots & & & \ddots & & \vdots \\ 0 & 0 & 0 & \cdots & 0 & 1 \\ c_n & c_{n-1} & c_{n-2} & \cdots & c_2 & c_1 \end{bmatrix},$$

is nonnegative and, of course, has eigenvalues $\lambda_1, \lambda_2, \ldots, \lambda_n$. ∎

Corollary 2.1. *Let $\sigma = (\lambda_1, \lambda_2, \ldots, \lambda_n)$ be a real n-tuple, where*

$$\lambda_1 \geq \lambda_2 \geq \cdots \geq \lambda_p \geq 0 > \lambda_{p+1} \geq \cdots \geq \lambda_n,$$

and

$$\lambda_1 + \lambda_{p+1} + \cdots + \lambda_n \geq 0.$$

Then there exists a nonnegative matrix with spectrum σ.

Proof. Use the method in the proof of Theorem 2.3 to construct a nonnegative $(n-p+1) \times (n-p+1)$ matrix B with eigenvalues $\lambda_1, \lambda_{p+1}, \ldots, \lambda_n$. Then the nonnegative matrix

$$B \dotplus \operatorname{diag}(\lambda_2, \lambda_3, \ldots, \lambda_p)$$

has spectrum σ. ∎

Corollary 2.1 implies the following result.

Corollary 2.2 (Perfect [14]). *Let σ be a real n-tuple, and suppose that it is possible to partition σ into subsets in such a way that* (i) *each subset contains one or more nonnegative numbers, and* (ii) *the sum of the negative numbers (if any) in each subset does not exceed the largest positive number. Then σ is the spectrum of a nonnegative matrix.*

7.3. SIMILARITY TO NONNEGATIVE MATRICES AND TO DOUBLY STOCHASTIC MATRICES

In this section we consider the ultimate question: What are necessary and sufficient conditions for a given matrix to be similar to a nonnegative matrix or to a doubly stochastic matrix; or equivalently, for the existence of a nonnegative matrix with prescribed elementary divisors? This problem includes the inverse spectrum problem which we discussed in the last section and which is unsolved. The two problems are equivalent if the prescribed eigenvalues are distinct.

We start with matrices of small orders. The inverse spectrum problem for a 2×2 matrix is very simple (Problems 17 and 18), and the inverse elementary divisor problem for such matrices is trivial. A complex 2×2 matrix can be similar to a nonnegative matrix if either it has two linear elementary divisors $\lambda - \lambda_1$ and $\lambda - \lambda_2$ with $\lambda_1 \geq |\lambda_2|$ (λ_2 being real), or a single quadratic elementary divisor $(\lambda - \lambda_1)^2$ with $\lambda_1 \geq 0$. In the first case the matrix is similar to a nonnegative matrix as in Problem 17. In the second case it is similar to the nonnegative matrix

$$\begin{bmatrix} \lambda_1 & 1 \\ 0 & \lambda_1 \end{bmatrix}.$$

The 3×3 matrices are the largest for which the inverse spectrum problem has been completely solved (Theorem 2.2). The corresponding inverse elementary divisor problem is quite straightforward.

Theorem 3.1. *If the spectrum* $\sigma = (\lambda_1, \lambda_2, \lambda_3)$ *of a complex* 3×3 *matrix satisfies the conditions of Theorem* 2.2, *then the matrix is similar to a nonnegative matrix.*

Proof. Let C be the given complex matrix. If C has three linear elementary divisors, then C is similar to one of the nonnegative matrices constructed in the proof of Theorem 2.2. Suppose now that the elementary divisors of C are $\lambda - \lambda_1$ and $(\lambda - \lambda_2)^2$. If λ_2 is nonnegative, then the Jordan normal form of C is nonnegative. It remains to consider the case where $\lambda_2 = -\alpha < 0$, $\lambda_1 = r \geq 2\alpha > 0$, and the elementary divisors of C are $\lambda - r$ and $(\lambda + \alpha)^2$. But then C is similar to the nonnegative matrix

$$\frac{1}{3}\begin{bmatrix} r - 2\alpha & 2r + 2\alpha & 0 \\ r + \alpha & r - 2\alpha & r + \alpha \\ r + \alpha & r + \alpha & r - 2\alpha \end{bmatrix}. \quad \blacksquare$$

For larger matrices we can only try to solve the inverse elementary divisor problem modulo the inverse spectrum problem: Given a nonnegative matrix A

with spectrum σ, does there exist a nonnegative matrix with spectrum σ and with arbitrarily prescribed elementary divisors, provided that elementary divisors corresponding to nonreal eigenvalues occur in conjugate pairs? We answer the question in the affirmative in the case when the given nonnegative matrix is a diagonalizable positive matrix. We require the following auxiliary result.

Lemma 3.1. *Let S be an $n \times n$ nonsingular complex matrix whose first p columns are real, and the remaining columns are not real and consist of $q = (n - p)/2$ pairs of complex conjugate n-tuples: columns t and $q + t$ being conjugate for $t = p + 1, p + 2, \dots, p + q$. Then*

(a) *the first p rows of S^{-1} are real, and*

(b) *rows t and $q + t$ of S^{-1} are conjugate for $t = p + 1, p + 2, \dots, p + q$.*

Proof. (a) Suppose that the conjugate columns t and $q + t$ of S are transposed, $t = p + 1, p + 2, \dots, p + q$. The resulting matrix is $\overline{S}$, the conjugate of S. Hence

$$\det(S) = (-1)^q \overline{\det(S)}\ .$$

Thus $\det(S)$ is real if q is even, and pure imaginary if q is odd. Similarly, if $1 \le h \le p$, then for any k,

$$\det(S(k|h)) = (-1)^q \overline{\det(S(k|h))}\ ,$$

and therefore $\det(S(k|h))/\det(S)$ is real. Hence if $1 \le h \le p$, then

$$(S^{-1})_{hk} = (-1)^{h+k}\det(S(k|h))/\det(S)$$

is real for $k = 1, 2, \dots, n$, and thus the first p rows of S^{-1} are real.

(b) Let $S = (s_{ij})$ and $S^{-1} = (z_{ij})$. Then

$$z_{tk} = (-1)^{t+k}\det(S(k|t))/\det(S),$$

and expanding $\det(S(k|t))$ by column $q + t$ we obtain

$$z_{tk} = (-1)^{t+k} \sum_{\substack{h=1 \\ h \neq k}}^{n} (-1)^{\eta+q+t-1} s_{h,q+t} \det(S(h, k|t, q + t))/\det(S), \quad (1)$$

where

$$\eta = \begin{cases} h, & \text{if } h < k, \\ h - 1 & \text{if } h > k. \end{cases}$$

If $p < t \le p + q$ and $h \ne k$, then the first p columns of the $(n - 2) \times (n - 2)$ matrix $S(h, k|t, q + t)$ are real, and the other columns are not real and consist of $q-1$ conjugate pairs. It follows, as in part (a), that

$$\det(S(h,k|t,q+t)) = (-1)^{q-1}\overline{\det(S(h,k|t,q+t))}\,,$$

and therefore $\det(S(h,k|t,q+t))$ is real or pure imaginary according as q is odd or even. Hence if $p < t \le p+q$ and $h \ne k$, then

$$\det(S(h,k|t,q+t))/\det(S)$$

is pure imaginary, and thus

$$\overline{\det(S(h,k|t,q+t))/\det(S)} = -\det(S(h,k|t,q+t))/\det(S). \quad (2)$$

Also,

$$\overline{s_{h,q+t}} = s_{ht}. \quad (3)$$

Therefore, using (1), (2), and (3),

$$\begin{aligned}\overline{z_{tk}} &= (-1)^{q+t+k}\sum_{\substack{h=1\\ h\ne k}}^{n}(-1)^{\eta+t}s_{ht}\det(S(h,k|t,q+t))/\det(S)\\ &= (-1)^{q+t+k}\det(S(k|q+t))/\det(S)\\ &= z_{q+t,k},\end{aligned}$$

for $t = p+1, p+2, \ldots, p+q$, and $k = 1,2,\ldots,n$. ∎

Theorem 3.2 (Minc [11]). *If there exists a diagonalizable positive matrix A with spectrum σ, then there also exists a positive matrix with spectrum σ and with arbitrarily prescribed elementary divisors, provided that elementary divisors corresponding to nonreal eigenvalues occur in conjugate pairs.*

Proof. Let $\sigma = (\lambda_1, \lambda_2, \ldots, \lambda_n)$, where $\lambda_1 > \lambda_2 \ge \cdots \ge \lambda_p$ are real, the remaining $n-p$ eigenvalues are nonreal, equal eigenvalues are consecutive, and λ_t is conjugate to λ_{q+t} for $t = p+1, p+2, \ldots, p+q$, where $q = (n-p)/2$. Let $Av_j = \lambda_j v_j$, $j = 1,2,\ldots,n$, where $v_1, v_2, \ldots, v_n$ are linearly independent, and v_t is conjugate to v_{q+t} for $t = p+1, p+2, \ldots, p+q$. If S is the $n \times n$ matrix whose jth column is v_j, $j = 1,2,\ldots,n$, then

$$S^{-1}AS = \operatorname{diag}(\lambda_1, \lambda_2, \ldots, \lambda_n) = D.$$

Let X be the subset of $\{2,3,\ldots,n-1\}$ so that $D + \Sigma_{t\in X}E_{t,t+1}$ is the Jordan normal form of the required matrix. We show that the matrix

$$B = (b_{ij}) = S\Bigl(\sum_{t\in X}E_{t,t+1}\Bigr)S^{-1}$$

is real. Let $S^{-1} = (z_{ij})$. Then

$$b_{ij} = \sum_{t \in X} s_{it} z_{t+1,j}, \qquad i, j = 1, 2, \ldots, n.$$

Now, s_{it} is real for $t = 1, 2, \ldots, p$, and, by Lemma 3.1(a), $z_{t+1,j}$ is real for $t = 1, 2, \ldots, p-1$. Furthermore, if $p < t < p + q$ and t belongs to X (note that p, $p + q$, n do not belong to X), then $t + q \in X$, and

$$\overline{s_{it}} = s_{i,t+q} \quad \text{and} \quad \overline{z_{t+1,j}} = z_{t+q+1,j},$$

for all i and j. Therefore

$$\sum_{\substack{t=p \\ t \in X}}^{n} s_{it} z_{t+1,j} = \sum_{\substack{t=p+1 \\ t \in X}}^{p+q-1} \left(s_{it} z_{t+1,j} + s_{i,t+q} z_{t+q+1,j} \right)$$

$$= \sum_{\substack{t=p+1 \\ t \in X}}^{p+q-1} 2 \operatorname{Re}\left(s_{it} z_{t+1,j} \right).$$

Hence b_{ij} is real for all i and j.

Now, $SDS^{-1} = A$ is positive, and therefore $A + \theta B$ is positive for sufficiently small positive θ. Thus

$$A + \theta B = S\left(D + \theta \sum_{t \in X} E_{t,t+1} \right) S^{-1}$$

is positive for such θ, and has the required elementary divisors. ■

The condition that the given matrix A be diagonalizable and positive are necessary for our proof. It is not known, however, whether the theorem holds without this assumption. Specifically, it is not known: (a) Whether for every nonnegative (or even positive) matrix there exists a diagonalizable nonnegative (positive) matrix with the same spectrum; (b) whether for every nonnegative diagonalizable matrix there exists a nonnegative matrix with the same spectrum but with arbitrarily prescribed elementary divisors, subject to the condition that elementary divisors corresponding to nonreal eigenvalues occur in conjugate pairs. In the case of doubly stochastic matrices, the answer to problem (b) is in the negative, as we shall show in the following example.

Example 3.1. Show that there exists a doubly stochastic 3×3 matrix with eigenvalues, $1, -\frac{1}{2}, -\frac{1}{2}$, but there is no doubly stochastic 3×3 matrix with elementary divisors $\lambda - 1$ and $(\lambda + \frac{1}{2})^2$.

The reader should have no difficulty in verifying that the doubly stochastic matrix

$$C = \frac{1}{2}\begin{bmatrix} 0 & 1 & 1 \\ 1 & 0 & 1 \\ 1 & 1 & 0 \end{bmatrix}$$

has eigenvalues $1, -\frac{1}{2}, -\frac{1}{2}$.

Suppose that $A = (a_{ij})$ is a doubly stochastic 3×3 matrix with elementary divisors $\lambda - 1$ and $(\lambda + \frac{1}{2})^2$. By Schur's triangularization theorem (see [9], Chapter I, 4.10.2), there exists an orthogonal matrix $S = (s_{ij})$ such that

$$S^{\mathrm{T}}AS = B = \begin{bmatrix} 1 & 0 & 0 \\ 0 & -\frac{1}{2} & a \\ 0 & 0 & -\frac{1}{2} \end{bmatrix},$$

where a is some nonzero real number. Since the first column of S is a multiple of $[1 \quad 1 \quad 1]^{\mathrm{T}}$, we can assume without a loss of generality that it is $[1 \quad 1 \quad 1]^{\mathrm{T}}/\sqrt{3}$. Then

$$\begin{aligned} a_{ii} &= (SBS^{\mathrm{T}})_{ii} \\ &= \tfrac{1}{3} - \tfrac{1}{2}s_{i2}^2 - \tfrac{1}{2}s_{i3}^2 + as_{i2}s_{i3} \\ &= as_{i2}s_{i3}, \qquad i = 1,2,3, \end{aligned}$$

since $\frac{1}{3} + s_{i2}^2 + s_{i3}^2 = \|S_{(i)}\|^2 = 1$ for $i = 1,2,3$. But the trace of the nonnegative matrix A is zero, and therefore $a_{ii} = 0$ for $i = 1,2,3$. Now, $a \neq 0$, and thus we must have

$$s_{12}s_{13} = s_{22}s_{23} = s_{32}s_{33} = 0.$$

But this is impossible, since an orthogonal 3×3 matrix without 0's in its first column cannot have two 0's in either of its other columns. ■

Observe that Example 3.1 implies that for any n, $n \geq 3$, there exists an $n \times n$ complex matrix B with the same spectrum as a given doubly stochastic matrix C, but not similar to any doubly stochastic matrix, even if the elementary divisors of B satisfy the usual restrictive conditions specified in Theorem 3.3 below (see Problem 25). This situation, however, cannot occur if the given doubly stochastic matrix happens to be positive and diagonalizable, in which case a result analogous to Theorem 3.2 also holds for doubly stochastic matrices.

Theorem 3.3 (Minc [11]). *Given a diagonalizable positive doubly stochastic matrix C, there exists a positive doubly stochastic matrix with the same spectrum as C, and with arbitrarily prescribed elementary divisors, provided that they do* *not include* $(\lambda - 1)^k$ *with* $k > 1$, *and that the prescribed elementary divisors corresponding to nonreal eigenvalues occur in conjugate pairs.*

Proof. Let $D = \operatorname{diag}(\lambda_1, \lambda_2, \ldots, \lambda_n)$, where $\lambda_1 = 1, \lambda_2, \ldots, \lambda_n$ are the eigenvalues of C arranged as in the proof of Theorem 3.2, and let $D + \sum_{t \in X} E_{t,t+1}$ be the Jordan normal form of the required doubly stochastic matrix. Let S be the nonsingular matrix whose jth column is an eigenvector corresponding to λ_j, $j = 1, 2, \ldots, n$, the first column being $[1 \quad 1 \quad \cdots \quad 1]^{\mathrm{T}}/\sqrt{n}$, columns $2, 3, \ldots, p$ (if any) being real, and columns j and $j + p$ (if $p < n$) being conjugate for $j = p + 1, p + 2, \ldots, q$, $q = (n - p)/2$. Then, as in the proof of Theorem 3.2, the matrix

$$B = S\left(D + \theta \sum_{t \in X} E_{t,t+1}\right)S^{-1}$$

is positive for sufficiently small positive θ, and has the required elementary divisors. It remains to show that B is doubly stochastic.

The first row of S^{-1} is $[1 \quad 1 \quad \cdots \quad 1]/\sqrt{n}$, since it is an eigenvector corresponding to 1 of positive doubly stochastic matrix A^{T}, and the product of this row and the first column of S is 1. It follows that all the other row sums of S^{-1} and all the column sums of S, except the first, are zero. Thus the first row of $S^{-1}J_n$ is $[1 \quad 1 \quad \cdots \quad 1]/\sqrt{n}$ and all its other rows are zero, and J_nS is the transpose of $S^{-1}J_n$. If we write $D + \theta\sum_{t \in X} E_{t,t+1}$ in the form $1 \dotplus G$, where G is $(n - 1) \times (n - 1)$, then

$$(1 \dotplus G)(S^{-1}J_n) = S^{-1}J_n,$$

and

$$J_n(1 \dotplus G) = J_nS.$$

It follows that

$$BJ_n = S\big((1 \dotplus G)S^{-1}J_n\big) = S\big(S^{-1}J_n\big) = J_n,$$

and similarly

$$J_nB = \big(J_nS(1 \dotplus G)\big)S^{-1} = (J_nS)S^{-1} = J_n. \quad \blacksquare$$

The condition in Theorem 3.3 that the given diagonalizable doubly stochastic matrix C be positive is essential for our proof. It is not known whether the theorem holds if C is a diagonalizable, irreducible, but not necessarily positive, $n \times n$ doubly stochastic matrix.

Example 3.2. Show that there exists an irreducible, diagonalizable, 4×4 nonpositive doubly stochastic matrix C, and a nondiagonalizable doubly stochastic matrix B with the same spectrum as C.

The reader can easily verify that the doubly stochastic matrices

$$C = \frac{1}{3}\begin{bmatrix} 0 & 1 & 1 & 1 \\ 1 & 0 & 1 & 1 \\ 1 & 1 & 0 & 1 \\ 1 & 1 & 1 & 0 \end{bmatrix}$$

and

$$B = \frac{1}{6}\begin{bmatrix} 0 & 2 & 2 & 2 \\ 2 & 0 & 2 & 2 \\ 3 & 1 & 0 & 2 \\ 1 & 3 & 2 & 0 \end{bmatrix}$$

have the same spectrum $(1, -\frac{1}{3}, -\frac{1}{3}, -\frac{1}{3})$. However, C is symmetric, and therefore diagonalizable, whereas the elementary divisors of B are $\lambda - 1$, $\lambda + \frac{1}{3}, (\lambda + \frac{1}{3})^2$, and thus B is not diagonalizable. ■

PROBLEMS

1 Find a nonnegative 2×2 matrix with eigenvalue -3.

2 Find a nonnegative 4×4 matrix with eigenvalue $3 - 5i$.

3 Show that there does not exist a stochastic matrix with eigenvalue $(3 + 4i)/5$.

4 For each of the following numbers find a doubly stochastic 4×4 matrix which has the number as one of its eigenvalues: (i) $\alpha + (1 - \alpha)i$; (ii) $\alpha - (1 - \alpha)i$; (iii) $-\alpha + (1 - \alpha)i$; (iv) $-\alpha - (1 - \alpha)i$; where α is a real number, $0 \le \alpha \le 1$.

5 Prove Corollary 1.2 using either Theorems 1.1 or 1.6.

6 Find a doubly stochastic 3×3 matrix with eigenvalue $(-1 + \sqrt{3}\,i)/4$.

7 Find a doubly stochastic 4×4 matrix with eigenvalue $(1 + 2i)/3$.

8 Show that $\alpha = (-5 + 4\sqrt{3}\,i)/10$ does not belong to Π_4. Find a doubly stochastic 4×4 matrix with eigenvalue α.

9 Use Theorem 1.6 to prove Theorem 1.4.

10 Use Theorem 1.6 to prove that the modulus of an eigenvalue of a stochastic matrix cannot exceed 1.

11 Verify that the arcs of the boundary of Θ_7 are given by the parametric equations in the following table [7]. (For notation see Theorem 1.8)

a'/b'	a''/b''	Equations
1/7	1/6	$\lambda^7 - t = (1-t)\lambda$
1/6	1/5	$\lambda^6 - t = (1-t)\lambda$
1/5	1/4	$\lambda^5 - t = (1-t)\lambda$
1/4	2/7	$\lambda^7 - t = (1-t)\lambda^3$
2/7	1/3	$\lambda(\lambda^3 - t)^3 = (1-t)^3$
1/3	2/5	$(\lambda^3 - t)^2 = (1-t)^2\lambda$
2/5	3/7	$\lambda^7 - t = (1-t)\lambda^2$
3/7	1/2	$\lambda(\lambda^2 - t)^3 = (1-t)^3$

12 Use Theorem 1.8 to determine the boundary of Θ_5. Sketch the graph of the region Θ_5.

13 Determine the boundary of Θ_6, and sketch it.

14 Show that any complex number is an eigenvalue of a positive 3×3 circulant.

15 Show that if $A \geq B \geq 0$ and $C \geq D \geq 0$, then $AC \geq BD$.

16 Show that if $A = (a_{ij})$ is an $n \times n$ nonnegative matrix, and $D = \operatorname{diag}(a_{11}, a_{22}, \ldots, a_{nn})$, then $A^m - D^m$ is nonnegative for any positive integer m.

17 Show that if $\sigma = (\lambda_1, \lambda_2)$ satisfies conditions (1) and (2), Section 7.2, then there exists a nonnegative matrix of the form

$$\begin{bmatrix} a & b \\ b & a \end{bmatrix},$$

with σ as its spectrum.

18 Show that for any real number λ_2, $|\lambda_2| \leq 1$, there exists a 2×2 doubly stochastic matrix with eigenvalue λ_2.

19 **(a)** Find an irreducible nonnegative 3×3 matrix with eigenvalues $4, 2, -1$.

(b) Find a direct sum of two nonnegative matrices that is similar to $\operatorname{diag}(4, 2, -1)$.

(c) Find a stochastic 3×3 matrix with eigenvalues $1, \frac{1}{2}, -\frac{1}{4}$.

20 Prove Corollary 2.2.

21 Find a nonnegative 5×5 matrix with eigenvalues $4, 2, 0, -1, -3$.

22 Find a stochastic 5×5 matrix with eigenvaues $1, \frac{1}{2}, 0, -\frac{1}{4}, -\frac{3}{4}$.

23 Find a nonnegative 3×3 matrix with eigenvalues $4, -\sqrt{3} + i, -\sqrt{3} - i$.

24 Find a nonnegative matrix similar to the matrix

$$A = \begin{bmatrix} 3 & 2 \\ -2 & -1 \end{bmatrix}.$$

Show that there is no stochastic matrix similar to A.

25 Find a diagonalizable $n \times n$ doubly stochastic matrix C, and a complex matrix B with the following properties:

(a) B and C have the same spectrum;

(b) B has no elementary divisors $(\lambda - 1)^k$, with $k > 1$;

(c) all elementary divisors of B corresponding to nonreal eigenvalues occur in conjugate pairs;

(d) B is not similar to any doubly stochastic matrix.

REFERENCES

1. A. Berman and R. J. Plemmons, *Nonnegative Matrices in the Mathematical Sciences*, Academic Press, New York, 1979.
2. N. Dmitriev and E. Dynkin, On the characteristic numbers of a stochastic matrix, *C. R. (Dokl.) Acad. Sci. URSS* **49** (1945), 159–162.
3. N. Dmitriev and E. Dynkin, On characteristic roots of stochastic matrices, *Izv. Akad. Nauk SSSR, Ser. Mat.* **10** (1946), 167–184 (in Russian; English translation in [17]).
4. M. Fiedler, Eigenvalues of nonnegative symmetric matrices, *Linear Algebra Appl.* **9** (1974), 119–142.
5. S. Friedland, On an inverse problem for nonnegative and eventually nonnegative matrices, *Israel J. Math.* **29** (1978), 43–60.
6. D. Hershkowits, *Existence of Matrices Satisfying Prescribed Conditions*, M.S. Thesis, Technion, Haifa, 1978.
7. F. I. Karpelevič, On the characteristic roots of matrices with nonnegative elements, *Izv. Akad. Nauk SSSR Ser. Mat.* **15** (1951), 361–383 (in Russian).
8. R. Loewy and D. London, A note on an inverse problem for nonnegative matrices, *Linear and Multilinear Algebra* **6** (1978), 83–90.
9. M. Marcus and H. Minc, *A Survey of Matrix Theory and Matrix Inequalities*, Allyn and Bacon, Boston, 1964.
10. H. Minc, Inverse elementary divisor problem for doubly stochastic matrices, *Linear and Multilinear Algebra* **11** (1982), 121–131.
11. H. Minc, Inverse elementary divisor problem for nonnegative matrices, *Proc. Amer. Math. Soc.* **83** (1981), 665–670.
12. L. Mirsky, Results and problems in the theory of doubly-stochastic matrices, *Z. Wahrsch. Verw. Gebiete* **1** (1963), 319–334.
13. L. Mirsky, Inequalities and existence theorems in the theory of matrices, *J. Math. Anal. Appl.* **9** (1964), 99–118.

14. H. Perfect, Methods of constructing certain stochastic matrices, *Duke Math. J.* **20** (1953), 395–404.
15. F. L. Salzmann, A note on eigenvalues of nonnegative matrices, *Linear Algebra Appl.* **5** (1972), 329–338.
16. H. R. Suleimanova, Stochastic matrices with real eigenvalues, *Dokl. Akad. Nauk SSSR* **66** (1949), 343–345 (in Russian; English translation in [17]).
17. J. Swift, *The Location of Characteristic Roots of Stochastic Matrices*, M.Sc. Thesis, McGill University, Montreal, 1972.

General References

1. A. Berman and R. J. Plemmons, *Nonnegative Matrices in the Mathematical Sciences*, Academic Press, New York, 1979.
2. F. R. Gantmacher, *The Theory of Matrices*, Chelsea, New York, 1959.
3. F. R. Gantmacker and M. G. Krein, *Oszillationsmatrizen*, *Oszillationskerne*, *und kleine Schwingungen mechanischer Systeme*, Akademie-Verlag, Berlin, 1960.
4. G. H. Hardy, J. E. Littlewood and G. Pólya, *Inequalities*, 2nd Edition, Cambridge University Press, London, 1952.
5. M. Marcus and H. Minc, *A Survey of Matrix Theory and Matrix Inequalities*, Allyn and Bacon, Boston, 1964 (or, Prindle, Weber and Schmidt, Boston, 1970).
6. H. Minc, *Permanents*, *Encyclopedia of Mathematics and Its Applications*, vol. 6, Addison-Wesley, Reading, Mass., 1978.

Index of Symbols

Index